Lubricants
and Their
Applications

Lubricants and Their Applications

Robert W. Miller
Jigsaw Services Co., Tempe, Arizona

McGraw-Hill, Inc.

New York   San Francisco   Washington, D.C.   Auckland   Bogotá
Caracas   Lisbon   London   Madrid   Mexico City   Milan
Montreal   New Delhi   San Juan   Singapore
Sydney   Tokyo   Toronto

Library of Congress Cataloging-in-Publication Data

Miller, Robert W.
  Lubricants and their applications / Robert W. Miller.
      p.   cm.
    ISBN 0-07-041992-2 (alk. paper)
    1. Lubrication and lubricants.   I. Title.
  TJ1075.M535   1993
  621.8′9—dc20   92-271
                                                   CIP

1 2 3 4 5 6 7 8 9 0   DOC/DOC   9 9 8 7 6 5 4 3

ISBN 0-07-041992-2

*The sponsoring editor for this book was Robert W. Hauserman, the editing
supervisor was Mitsy Kovacs, and the production supervisor was Donald
Schmidt. It was set in Palatino by McGraw-Hill's Professional Book Group
composition unit.*

Printed and bound by R. R. Donnelley & Sons Company.

This book is printed on recycled, acid-free paper
containing a minimum of 50% recycled de-inked
fiber.

Contents

Preface

This book will provide basic knowledge on the "magic" of lubricants and common industrial applications. The information will assist engineers and maintenance managers in the proper selection of lubricants, a supplier, and services that will contribute to the effectiveness of the overall maintenance program. Some major oil companies and distributors are currently staffing the field with young, inexperienced salespeople. It is vital that industries increase their knowledge of these basics to better improve machine availability, production, and profit.

There are many good educational and informative manuals published by oil and equipment manufacturers. Such technical manuals, along with the petroleum product data sheets and equipment operational manuals, can be very helpful in increasing performance in the field of lubrication and maintenance. This book will assist in the understanding of the lubrication information in those manuals and from the lubricant supplier. Although specifications, classifications, and designations will continue to change, the basic fundamentals will remain constant.

As a retired field manager who has spent over 34 years with two major oil companies servicing industrial and fleet accounts, I have sold and engineered the use of a wide variety of lubricants. My experience was gained by working in the field with technical and mechanical personnel and machinery of all types.

The information is presented in a straightforward, practical, and easily read manner and is free of technical line graphs and jargon. There is an appendix containing charts on various lubricant grading systems and a glossary of terms. All product names are registered trademarks by their respective manufacturer as are the qualification names of automatic transmission fluids.

Acknowledgments

I would like to express my deepest appreciation to the following individuals for their support, advice, and guidance in the preparation of this book: Steve Horrocks and Don Walls, Hasco Oil Co., Long Beach, California; Jim McCaslin, MacValley Oil Co., Oxnard, California; Joe Dufort, Wilcox & Flegel, Longview, Washington; and Jeanne Woolford, friend and proofreader.

Robert W. Miller

1
Supplier-Customer Relationships

Authority

The authority to select and purchase lubricants varies among companies. The authority may be vested as follows:

1. Purchasing department only
2. Maintenance department only
3. Engineering department only
4. Any combination thereof

Some purchasing departments want to maintain control over all purchasing because that is their job, while the engineering department wants to control which products will be purchased because they are closer to the technical aspects of lubrication. Maintenance wants to control the lubricants in the plant because they think they know what products work the best and which supplier can provide prompt product delivery and technical assistance.

Unfortunately, everyone is correct. Purchasing should have some control over the competitive pricing of your lubricants. But they can often can be obsessed with saving pennies on a product that is not multipurpose or cost-effective in use.

Engineering should be technically involved with products, particularly if the lubricant's application can affect production or have an adverse environmental impact. Today, just an adverse environmental impact can immediately disqualify a product from being used in a plant.

Maintenance sees the results from the use of the lubricants and knows that many times saving pennies on inadequate products costs big dollars in maintenance.

Many people consider the ideal concept to be an engineering-maintenance team, working with suppliers, to select the proper types of lubricants for the plant. A list of selected lubricants and proposed services can then be taken to purchasing with justification to purchase.

This can be a very good arrangement for an efficient maintenance program because the engineering-supplier-maintenance team will be better informed on the proper types of lubricants needed for the plant and those which serve multiple purposes. They will also be better informed of the quality of the service potential, both technical and operational, of each supplier. These elements should be a major part in the selection of a supplier and products.

The team should evaluate the following products and capabilities of each supplier:

1. What are the general classifications of products needed to lubricate the plant and are those lubricants suitable for multiple purposes? Multipurpose products will reduce the number of different products which the plant is required to stock.

2. What is the reputation of each lubricant manufacturer that supplies in the area and what services are available from the supply point and the manufacturer?

3. Do the lubricant manufacturer and supplier have services available which could help you in the lubrication program? Examples are lubricant recommendations for new machinery, lubrications charts, experienced personnel, laboratory services, on-site testing of some lubricants, and prompt delivery. These are the tools to build a better maintenance program.

4. Are containers returnable and do they carry a deposit, or are they nonreturnable? A typical industrial plant will normally return over 95 percent of their returnable drums. A construction job will normally return only about 60 percent of their returnable drums because the others are damaged, lost, or stolen.

5. Is there a suggested method for disposal of nonreturnable containers and waste oil?

6. Does your supplier have a prompt procedure for the correction of in-

voices which are in error? Are there delivery discounts and are the prices firm for a particular period of time?

Maybe good old Charlie with ABC Distributing is a nice guy, plays on your softball team, knows your wife, and is technically astute, but Ole Charlie is not going to be around forever. The engineering-maintenance team must look at all the factors which will contribute to the success of the lubrication program. This is why the authority should be vested in a team rather than one individual in your company. The team must choose the product lines, the supplier with the best track record, and the one which will be the most overall cost-effective for the company.

Cost-Effectiveness

To purchasing, cost-effectiveness normally means they were able to purchase something for a lower price. The real meaning of cost-effectiveness is the total cost of buying and handling a product and the resulting savings from using that product. Let's relate that definition to lubricants.

Although explaining cost-effectiveness is somewhat redundant, the real meaning can be easily overlooked in industry today.

The effective use of an oil, which cost $2 for 6 months, would have an annual oil cost of $4. If you were to purchase a product for $2.20 per gallon and use it effectively for 7 months, your annual oil cost would be only $3.77. That is a very simple way to explain cost-effectiveness, but there is far more to cost-effectiveness than oil cost and length of service.

To expand this simple comparison, let's add in the real, but less obvious, costs. Let's assume it costs $200 in labor and overhead to change out that oil system and it takes 100 gal of $2.20 per gallon premium oil to do so. Let's further assume that a machine generated $1000 profit per 8-hour day for the company and the change-out shuts that machine down for 1 hour. We are now looking at the real cost of the oil change as follows:

Labor and overhead to change the system	$200.00
1 hour of profit lost: $1000 divided by 8 hours	$125.00
Handling costs for 2 drums of oil	$ 40.00
Premium oil cost: 100 gal × $2.20 per gal	$220.00
Total costs	$585.00

The real cost of changing that 100-gal oil system is $5.85 per gallon for the premium oil, which would last 7 months, compared with $5.65 per gallon ($0.20 per gallon price difference) of the lower-quality oil, which would last only 6 months.

If, by purchasing a better-quality product, the above costs can be incurred over 7 months' service rather than 6 months with a lesser-quality product, there will be an annual savings of $127 using the premium product (total cost to service the system divided by months of service multiplied by 12 months).

The savings of $127 per year relates directly to a reduction in the real cost of the premium oil of *almost $0.74 per gallon,* making the premium product less expensive to buy than the lesser-quality product by *$0.54 per gallon.*

That is cost-effectiveness!

Naturally there are applications in which quality products are not cost-effective, but a little imagination can be used here. For example, a premium hydraulic oil, having served its primary function, can be inexpensively filtered and used as a chain or squirt-can lubricant. Quality products are normally cost-effective.

Few maintenance groups are overstaffed. With the use of quality products and a good maintenance program, they can spend more time on improving machine production time and profitability and less time on routine matters and emergencies.

Customer-Supplier Relations

Your relationship with your oil supplier can take three directions:

1. *Adversary:*   Browbeat potential suppliers to bid low. Select the lowest bidder on the products of unknown performance or minimum specifications and then demand services be provided that eliminate any small profit the supplier might have gained.

2. *Customer-dominated:*   Clearly state exactly the products needed, get your bids, and run your lubrication program without outside assistance or fresh ideas. This system perpetuates mediocrity.

3. *Customer-supplier team:*   Start with a supplier that provides quality products with equitable pricing and the technical services to obtain the maximum utilization of those quality products. Team up with a supplier who can and will provide the services necessary to assist

you in the completion of your responsibilities. You can then work together as a "maintenance team."

Many companies have found the "customer-supplier team" to be the most satisfactory and rewarding. The team should meet every 2 or 3 months to discuss the following areas of common interest:

1. Consolidation of products to reduce the number of lubricants necessary in the plant. Specialized products might be more cost-effective.
2. Potential of bulk deliveries to reduce product and handling costs. Could bulk high-volume products be piped into centralized locations in the plant for ease of distribution and reduced handling expense?
3. Which products can be filtered and reused for less critical applications or perhaps burned as fuel in a boiler? Reclamation and reuse of products can greatly add to the cost-effectiveness of any product.
4. What oil testing can be done, both on-site and laboratory, which would allow product life and performance to be extended?
5. Can your supplier conduct educational clinics to familiarize the maintenance mechanics or oilers with the products in the plant? Can they provide management with a list of emergency product substitutions should a critical product be in short supply in the plant?
6. Names and home phone numbers in case of a plant emergency.

It is easy to see how the team concept can be beneficial and profitable to all concerned parties. Customers have outside assistance in satisfying their areas of responsibility. Their lubrication and maintenance programs and personnel can be made more efficient and knowledgeable. Machine availability and profit increase.

Suppliers are fully aware of their obligations to the customer and have a clear path of communication to resolve problems and present new ideas to assist the customer.

The supplier can be requested to put new ideas and results of tests in writing to document your joint efforts and results. This is the information which will justify your team concept and supplier choice to the purchasing department and upper management.

The team concept is a "win-win" situation for both the customer and the supplier.

Customers are assured of getting quality multipurpose lubricants, delivered promptly, in the most efficient container size with correct or correctable billing, and have a source of qualified technical assistance and emergency deliveries as needed.

Suppliers have a good solid customer purchasing on a regular basis, and are profiting enough to allow emergency deliveries and technical assistance on a prompt basis. They can also regulate their stock to ensure product availability.

It is truly the best of two worlds and will contribute to an efficient maintenance system which will increase machine productivity, reduce maintenance costs, and increase profit.

2
Principles of Lubrication

Friction

Friction is an element of many faces. It is bad when it causes heat, wear, and reduced energy in a machine. It is good when it keeps our feet on the pavement as we walk and our car on the road as we drive down the highway. Friction is what holds nails in a board, allows our brakes to stop our car, and allows us to walk on ice, carefully of course, because if that small friction is overcome, we fall.

Two surfaces, moving against each other, have friction. Two rough metal blocks have a great deal of friction unless a lubricating fluid separates the two blocks to reduce the friction. That is the function of a lubricant or lubricating fluid. The lubricant can be mineral oil, grease, a soft metal or mineral, water, or similar substances. With the lubricating fluid in place, the two metal surfaces are separated, and although there is some movement between the lubricating fluid and the metal, the majority of movement is within the fluid itself. A good example of this is wet ice. On dry ice, the surface of the ice is not too slippery. As the surface of that ice starts to melt and a thin water phase is between the ice and your shoe, the ice will be very slippery. The water phase is the lubricating fluid between the ice and the shoe. Needless to say, the roughness of the two mating surfaces determines the friction between them. The friction between two rough-sawn metal surfaces is far greater than between two metal surfaces which have been mechanically or electrolytically polished. Even the friction between the finely polished surfaces is reduced by adding a lubricating fluid between them.

With lubricating mineral oils, the majority of the movement is within the oil structure, much the same as it was within the water on the ice. An oil that is too thin, however, allows the metal surfaces to touch and increase the friction. An oil that is too viscous separates the metal surfaces, but because the oil is so viscous, the internal friction within the oil layers is high. Compromise is the name of the game here. You need an oil that is thin enough to have a low internal friction yet heavy enough to separate the metal surfaces. If you can get a fluid, like a synthetic, which can separate the metal surfaces and still have low internal friction because of its uniform molecular structure, you have made progress in reducing the friction in the application.

In a grease application, as noted above, the grease separating the two metal surfaces reduces the friction between the surfaces. As the grease film is worn away, the frequency of metal-to-metal contact and the friction will increase. Even when the grease film is completely worn away, there will still be less friction than there would be on clean metal-to-metal surfaces. Some grease will remain in the valleys of the metal surfaces which will still be performing a friction-reducing function.

A soft powdered metal or mineral acts basically as the lubricating fluid. The internal friction within the structure of the powdered metal or mineral is far less than that of the two metal surfaces in contact. That is the reason, in certain applications, powdered zinc may be added to a grease as an antiseize agent or graphite or molybdenum disulfide (moly) might be added in an effort to reduce friction. They assist the lubricant carrier in the reduction of friction should their properties be needed. These fillers have a lattice structure and the planes of the structure slide across each other more easily than the structure slides across the metal surface. This action is similar to sliding a deck of cards apart. The top card sticks to your hand while the bottom card sticks to the table.

The graphite and moly differ in that water or some volatile material is needed to help split the layers of the structure in graphite but this is not necessary with moly. The layers of moly slide across each other more easily without outside assistance. Polytetrafluoroethylene (PTFE) is noted for its chemical inertness because of the strong carbon-fluorine bonds in its structure. Its structure shows very little resistance to sliding within the structure and very little tendency to form any strong bonds between itself and other materials. PTFE shows a very low friction coefficient in high-load and low-sliding-speed applications. The high wear rates of PTFE in some applications can be reduced by the introduction of a light fluid of lubrication.

Chemically active additives, which combine and change the structure of the mating metal surface, also affect the friction and wear of some applications. With the common extreme-pressure additives containing sulfur, phosphorus, or sometimes even chlorine compounds, those elements combine with a steel surface to form a minute film of iron sulfide or other combination depending on the element used. The structure of the iron sulfide has a lower shear strength than does the original metal. This lower shear strength allows metal-to-metal contact without the ripping and tearing involved with the original metal surfaces, should they come within contact. With the altered metal surface, there is almost a polishing action as a result of contact. Applications involving these types of additives must be evaluated for speed and loading to select the correct type of additive to be used to reduce the friction and wear.

Chlorine is probably the least used in lubricating applications owing to its potential for corrosion and promotion of oil oxidation and hydrolyzing tendency which can cause rust on ferrous metals. Chlorine films also tend to lose their lubricating effectiveness faster than sulfur and phosphorus films. Chlorine, combined with sulfur, is commonly used in cutting fluids to change the metal surface being worked and to reduce the friction between the metal being worked and the tool. At certain temperatures, chlorine is more effective in cutting fluids than is sulfur because they activate at different temperatures. They are extremely effective in reducing friction and improving surface finish in many heavy-duty machining operations.

Other factors contribute to changes in friction between two surfaces, but the factors mentioned are the most important in the field of lubrication.

Introduction to Lubrication

The first principle, also called the *right* principle of lubrication, is to put *the right lubricant in the right place at the right time.*

Lubrication-wise, a machine is very dumb! If you use the right oil in the right place at the right time, that machine will be happy. The job of the customer-supplier team is to provide the right oil!

The right oil is determined by many factors.

1. The type of lubrication in the application.

2. The manufacturer's recommendations. These should be followed so that all warranties will remain in effect. They are the most important, since manufacturers know their machines better than anyone else.

3. The viscosity and correct additives in the oil necessary to meet the machine's needs.

4. The ambient and operating temperatures of the application.

5. The loading or abnormal pressures and speed of the applications if changed from the original design.

The customer-supplier team should evaluate the operational conditions of the equipment to determine if abnormal conditions suggest a change in the lubricant recommendation to provide better lubrication.

An example of this type of evaluation occurred when a customer was using a series of large stamping presses. The plain bearings were lubricated by an automatic dispensing unit on the side of the press which pumped the oil up to the bearings on a 2-minute cycle. When the press was shut down for the weekend, the pump had to continue to pump oil to prevent a dry start on Monday morning. An oiler had to be on duty to prevent the pump reservoir from running dry. On Monday morning, the oil, pumped over the weekend, was all over the machine and on the floor and had to be cleaned up. Bearing life was only 3 months.

The supplier suggested that an NLGI 000 grade of lithium semifluid grease would stay in the bearing better than the recommended oil. The customer's aggressive equipment engineer set up a test.

The results of the testing proved the semifluid grease would (1) stay in the plain bearings better, (2) allow for a reduced lubrication cycle, (3) not drain back into the lubricator, which eliminated the need for the weekend oiler, (4) allow the engineer to tighten the bearing clearances, speed up the machine cycle, and increase productivity and bearing life.

As the end result of the test and changing the type of lubricant, the customer reduced maintenance costs, improved bearing life, and increased production. That customer, in that one department, saved money and increased gross profit enough to pay for all lubricants used in the entire plant for a year.

Oil performs several different functions. It cools or removes heat, can help seal against contaminants entering the machinery, reduces friction, and separates moving parts with a protective oil film. The right lubricant in the right place at the right time is vital to an effective maintenance program.

Let's take a look at all these factors.

Types of Lubrication

All metal surfaces, regardless of how finely finished, appear as a series of peaks and valleys upon close examination. The object of lubrication is to separate these peaks so contact is avoided and wear greatly reduced or eliminated. If we can successfully avoid that contact by keeping a full film of lubricant between the metal surfaces, that is known as full fluid film lubrication. This fluid film minimizes friction contact between the metal or bearing surfaces and consists of multiple layers of oil molecules. The layer of molecules in contact with the bearing surface maintains contact by "adhesion" while the different layers of oil molecules are held together by "cohesion." As the bearing surfaces move against each other, the molecule layers move slowly against the surface and more rapidly against each other, somewhat like sliding a deck of cards apart. The resistance between the layers of molecules is known as "fluid friction." In applications where an oil is used which is too viscous, the internal fluid friction can produce enough heat to shorten the life of the bearing surfaces.

Although a full fluid film is ideal, it rarely occurs in practice. More common is a condition called "boundary lubrication" where the bearing surfaces are primarily separated but the peaks of the bearing surfaces make some contact. This condition is very common upon the start-up and the shutdown of machinery. As loading on bearing surfaces increases more and more contact occurs and welding and tearing free of the peaks increases wear. The effects of this adverse condition can be minimized by the use of the proper antiwear or extreme-pressure additives. In gear sets the speed, loading, temperature, and type of gearing dictate the proper lubricant to use for maximum protection of the components.

Another condition would be in a bearing lubricated daily with a squirt oil can. When the oil is applied, a full fluid film would probably exist. As the bearing continued to turn, the oil film would gradually be reduced until a boundary lubrication condition existed and then finally an insufficient film of lubricant would allow a high wear rate. If that oil cup and squirt can could be replaced by an oil cup that was filled every day or a spring-loaded grease cup, that bearing would be assured of a continuous supply of lubricant throughout the day, and wear would be minimized.

The correct lubricant separates the moving parts and reduces friction between the parts and the internal friction within the fluid. It is now easy to see why the right lubricant in the right place at the right time is so very important to proper lubrication and good life of the components being lubricated.

Manufacturer's Recommendation

In the manufacturer's maintenance or operational manual, there is usually a listing of recommended lubricants and their points of application.

They are listed by lubricant type or the brand name of one or more of the major oil companies. This is an excellent basic guide to follow.

Some equipment representatives will voice an opinion on their preference, but warranties are honored on the equipment manufacturer's recommendations.

Most major oil companies publish some type of listing which shows major manufacturers' equipment and the oil companies' recommended lubricant to meet the equipment's needs. Most oil companies check these recommendations out with the manufacturer prior to publication. This is the second best method of choosing the correct type and viscosity of lubricant for the application from that supplier. Should you still have questions, most major oil companies can assist you via 800 phone numbers designed to provide product and equipment information.

Often the plant will already be using a product which could fulfill the needs of a new machine. This product may not be the best product for the job, but it will be a starting point for you. If any modifications have been made to the machine, normal operating conditions, or lubrication systems, those changes may affect the selection of the correct lubricant to be used.

Important: The product currently in use may be the wrong type of product! Do some research to confirm that it is the correct type and viscosity of product for the new machine. If the plant has abnormal operating conditions (such as high or low temperatures or shock loads) and/or you are experiencing short component life, you could be using the wrong type or viscosity of product under those operating conditions.

Viscosity and Additives

The correct viscosity of the lubricant is vital for proper lubrication. Viscosity is a measurement of the "resistance to flow," or how thick the oil is at a given temperature. The viscosity *must* be adequate to separate the moving parts at the operating temperature of the machine. An oil of the correct viscosity will do the job, but the proper additives can enhance the oil's performance.

If the application is a heavily loaded gear set, an extreme-pressure additive could be mandatory to protect the gear teeth. Since temperatures will elevate in an operating gear set, the addition of oxidation inhibitors will be helpful for longer oil life. Because some condensation is normal, owing to the effects of heating and cooling, a rust-inhibitor additive in the oil would be beneficial.

If the application is a hydraulic system, the extreme-pressure additive is not needed but an antiwear additive will extend the life of high-pressure hydraulic components. Rust and oxidation inhibitors also improve the performance of an oil in this application.

Again, the right viscosity and additive combination is vital to achieve correct lubrication. This is true whether the lubricant is 3-in-one oil or a heavy worm gear lubricant. There is more detailed information on viscosity and additives in the chapters that follow.

Temperature

When equipment manufacturers recommend a particular viscosity oil, they compute the typical normal temperature plus the normal temperature rise to arrive at the viscosity recommended. It is common knowledge that oil becomes thinner or less viscous as it increases in temperature and gets thicker or more viscous when it cools. If that machine is being used in a cold shop in Minnesota or a non-air-conditioned shop in Arizona, those temperatures will vary from the normal and the viscosity necessary will also vary.

If the ambient temperature is cold, a thinner oil is needed. This will permit the machine to start but will still provide the correct viscosity when the machine achieves the lower than normal operating temperature.

> THE LOWER THE TEMPERATURE,THE LIGHTER THE OIL.

If the ambient temperature is hot, you would use a more viscous oil so the machine, at the operating temperature, again has the correct viscosity of oil.

> THE HIGHER THE TEMPERATURE, THE HEAVIER THE OIL.

Conditions could be considered normal if that machine in Minnesota has an oil heater and the machine in Arizona has a heat exchanger to keep the oil within the correct temperature range for proper operation. In a lubricating system, which is not heated or cooled, the "ambient" or room and location temperature determine the normal for that operation.

The "normal" is the basis for the manufacturer's oil recommendation.

Loads and Pressures

The "KISS" method (Keep It Simple, Stupid) is a good way to explain loads and extreme pressures.

Take the common garden wheelbarrow, for example. The wheel bearing is slow-moving and not subject to high temperatures. Push the wheelbarrow along empty and there is very little "load" or "pressure" on the wheel bearing. Let's assume a "normal viscosity" oil will keep the shaft separated from the bearing with an adequate oil film.

Fill the wheelbarrow with wet sand, and the load on the bearing is increased. The normal oil cannot keep the shaft and the bearing separated with this increased load. A heavier or a more viscous oil is needed to form the proper oil wedge and to separate the parts.

Take that wheelbarrow full of wet sand and bounce it down a flight of stairs. "Shock loading" or "extreme pressure" has now been added to the bearing. Even the heavier oil film could now rupture and cause excessive bearing wear. Extreme-pressure additives would assist in preventing that bearing wear.

Remember this illustration of load, shock loading, and extreme pressure. How to cope with these and other adverse conditions is covered more fully in the application section.

THE HIGHER THE LOAD, THE HEAVIER THE OIL.

THE LIGHTER THE LOAD, THE LIGHTER THE OIL.

Speed of an Application

Take that same wheelbarrow with its shaft and plain sleeve bearing and add speed. Let's further assume it is designed to rotate at 1000 rpm (revolutions per minute) and it has a normal load. Further assume the correct viscosity oil is in the bearing for that operation.

As the shaft rotates, at *normal* speed, the correct viscosity oil is drawn or dragged around on the shaft to form an oil wedge between the shaft and the bearing (load zone of the bearing). This oil wedge actually lifts the shaft off the bearing surface, preventing metal-to-metal contact. If we use an oil which is too thick or too thin, an incorrect amount of oil is dragged into the load zone of the bearing.

Increasing the speed pulls too much of the normal oil into the bearing so a less viscous oil should be used to form the proper oil wedge. With the proper oil wedge in the load zone, the bearing is protected. Increased speed needs a lighter oil.

Decreasing the speed of the wheel, the oil wedge of the thin oil would be too thin in the load zone to protect the bearing. Therefore, a heavy or more viscous oil with more cohesion is needed to separate and protect the bearing surfaces with the proper oil wedge. Decreasing the speed changes the bearing's needs, and a heavier oil is required.

Thus the relationship between oil viscosity, speed, and the oil wedge is established.

> THE FASTER THE SPEED, THE LIGHTER THE OIL.
>
> THE SLOWER THE SPEED, THE HEAVIER THE OIL.

Modern lubricants are a balance of the correct viscosity oil and additives to enhance the performance of the oil under adverse operating conditions.

Oil Degradation

Oil in service is chemically and physically contaminated depending on the length of service and the conditions imposed on it by its application. Although they are linked together, in many cases, and one factor may influence another factor, we will try to separate them for ease of understanding.

A mineral oil lubricant in service will chemically change during that period of service by its exposure to heat, oxygen, the formation of acids, and the catalytic metals it comes in contact with in the application. Oxidation, due to high temperatures and extended service, will tend to darken the color of the oil, increase its viscosity, and make it take on an odor. Insoluble resins will ultimately be formed and become deposits forming in the application. The chemical structure of the oil will change, gradually turning to organic acids which are then corrosive and harmful to any equipment. Although additives will retard this chemical change, it will ultimately occur as the additives are depleted.

Contamination from the applications can make a major contribution to the degradation of the oil, both physically and chemically. Dirt can build up in a system, causing accelerated wear. Wear metals can act as a catalyst in the oxidation process as can the acids formed by the oxidation of the oil. Products of oxidation will act as a catalyst for further oxidation.

In a gasoline engine, the contamination comes from many sources. There might be fuel dilution from cold starts, short trips, and low engine temperatures. There will be by-products of poor combustion or an incorrect fuel-to-air ratio and blowby past the rings in a worn engine. Again, the wear metals, both normal and abnormal amounts, will contribute to the physical contamination as well as act as catalysts for more rapid oxidation.

A leaking head gasket can put a combination of water and antifreeze into the lubricating system. The presence of water has been known to leach out some of the zinc antiscuff additives. If the engine reaches and holds operating temperatures for a period of time, the water will vaporize off but will leave minerals behind in the oil. Glycol will oxidize rapidly with heat and form deposits in the oil and application as well as promote further oil oxidation. These elements will form acids or corrosive materials which will attach the soft bearing metals as well as the harder metals in an engine.

In a diesel engine, many of the same problems exist with these exceptions. Soot is a normal by-product of a diesel engine and will increase the viscosity of the oil. Soot is also thought to become abrasive as its level increases in an oil. Diesel fuel dilution is not uncommon in a diesel. Lab analysis of an oil sample with both soot (which thickens the oil) and fuel dilution (which thins the oil) could show the correct viscosity. Even with the correct viscosity indicated, the soot could be at an abrasive level and the fuel dilution will reduce the film strength of the oil to a point where accelerated wear occurs. The flash point of the oil will also be reduced by the fuel dilution.

Since diesel fuel contains more sulfur than gasoline, a diesel engine oil has to also fight the formation of acid from the fuel. Although the sulfur content of diesel in the United States is very good, it varies somewhat across the country depending on the crudes used and government requirements. Typically, a diesel engine oil will have a high TBN (total base number) to assist it in combating the acids formed. Some research has indicated that a higher TBN oil will reduce wear even in engines not subject to high-sulfur fuels. Not only does a higher TBN oil seem better in combating corrosive wear, but the oil property called "TBN retention" is an important factor. The additive composition will normally reflect in both the TBN and TBN retention. TBN is an antacid for an engine like an antacid tablet for a person. While some additive packages will show a good TBN number, like 10 to 14, their TBN retention might be poor and deplete more rapidly than will another similar oil. All these factors contribute to the degradation of the engine oil.

It is easy to see that the engine oil, in both a gasoline and a diesel engine, is the lifeblood of that engine. It has to fight internal and external contamination as well as do its original job of sealing, cooling, and lubricating the engine.

Compared with an engine oil, an industrial circulating oil has it made. The operating temperatures are normally closely controlled for long life, the small amount of physical contaminants are quickly and finely filtered out, wear metals are minimal, and there would be little chance for contamination from other products like diesel fuel or gasoline. There is a chance for water contamination from a leaking heat exchanger or water used in the manufacturing process. However, the recommended oil would be chosen for its ability to separate from water quickly which an engine oil cannot do because of the detergent-dispersant additive package. The water can then be drained off, leaving the circulating oil relatively free from moisture.

An industrial circulating oil has to contend with wear metals acting as catalysts, hot spots in the system, dirt entering the system, and a potential water leak on occasion. If the circulating system is overfilled or there is an air leak on the suction side of the circulating pump, air can be mixed with the oil and foaming can occur. Foaming can also occur if the oil return line allows the oil to drop into the oil in the reservoir. The drop and splashing will contribute to foaming. Not only does the air-oil mix make a very poor lubricant, it will promote more rapid oxidation in the oil. Any condition in a circulating system that is promoting foam should be immediately corrected. The vast majority of foaming conditions are caused by a mechanical defect or condition and not by a deficiency in the oil. If the oil is contaminated with some other product, foaming can occur but will be eliminated by changing out the oil. The exception to this might be the residual presence of a rust-preventive product on the components of the circulating system.

A unique situation was encountered in a steel mill in the middle west several years ago. A large melting pot turning gear, approximately 15 feet in diameter, was lubricated by a bath of gear oil. To keep it fluid in the cold months, oil heaters had been installed. The gear oil in the bath keep getting thinner and thinner the longer it remained in service. Laboratory analysis confirmed that solvent was being added to the gear oil. The fill caps and doors were sealed and locked to stop the addition of any foreign substance, but the thinning of the oil continued. The bath was drained and the oil bath reservoir was inspected. Almost a foot of granular carbon was found on the bottom of the reservoir, and the oil heaters, which were placed toward the bottom, were totally covered with carbon. The oil heaters were examined and found to be of a very high wattage per square inch of surface, much higher than is recommended for heating oil. Putting all the facts together, the problem was the heaters. Being placed so low in the tank and being of such high wattage, the heaters were actually cooking, or cracking, the gear oil. In the cracking, the light ends vaporized but were condensed in the upper part of the oil bath. This cracked portion of the gear oil was very similar to solvent. This solvent reduced the viscosity of the gear oil while the heavy carbonized ends settled to the bottom of the reservoir. The heaters were replaced with units of the proper wattage and the problem was solved.

It is hard to imagine everything possible that could go wrong and cause a problem. The key to solving a problem is "Never assume anything!" Check every avenue even remotely connected to the problem, the product, and the application. Oil analysis can be an excellent problem solver.

An oil in a steam turbine, which has a normal service life of well over 20 to 25 years, only has to contend with gradual oxidation caused by old age and the small catalytic effect of some wear metals which are normally quickly removed by filtration. Small quantities of water can and should be quickly separated out and removed. Water must be removed quickly to protect the equipment both from rust and from the depletion of the antirust inhibitors in the turbine oil. Some turbines develop a hot spot in the circulating system which will contribute to the depletion of the oxidations inhibitors. This will be a slow process and should be quickly picked up on through the routine oil analysis program which every power plant has, or should have, in its preventive maintenance program. The oxidation inhibitors in quality steam turbine oils are not designed for extreme conditions and will be rapidly vaporized off at elevated temperatures. Many times, a "gas" turbine oil utilizes different oxidation inhibitors which are designed for higher-temperature applications.

Gear oil normally oxidizes and gets contaminated with metal fines from the gear teeth. The extreme-pressure additives are rarely depleted. The continual addition of metal fines will act as a catalyst which, along with the agitation of the gear oil, promotes oxidation. Further oxidation is promoted by the high spot temperatures generated as the gear teeth mesh. The overall oxidation rate of a gear oil may be reduced by lowering a good-sized magnet down into the oil bath and allowing it to remain there while the oil circulates or is in motion. This will remove many of the ferrous fines from the oil, reducing the catalytic effect of the metal and extending the life of the bearings and gears lubricated by the oil. The magnet should be of good size and weight so it does not get drawn in to mesh with the gear teeth and cause a major failure.

When a gear case has an oil change, every effort should be made to clean out the deposits on the bottom of the case to ensure minimum contamination of the new oil. Magnetic drain plugs are also useful in keeping the amount of metal fines to a minimum.

Air compressors suffer from both oxidation and contamination. Any oil, mixed with the hot compressed air, will have a high oxidation rate. This oxidized oil can collect on the compressor valves and move downstream in the air supply lines and coolers. Naphthenic-type oils are sometimes recommended because they leave a lighter, fluffier carbon deposit than will a paraffinic-type oil. This is one reason, along with better oxidation resistance, that many are switching to the synthetic fluids for better performance. In most cases, the changeover is cost-effective in both maintenance and performance.

Self-Quiz

Mark your answers to the questions and compare them with the answers at the end of the book.

1. The equipment manufacturer's manual is an excellent source for the correct lubricants to use on a piece of equipment.
   a. True
   b. False
2. Probably the most important feature of the correct lubricant would be the:
   a. Additives
   b. Viscosity
   c. Brand name
3. If an equipment manufacturer recommends an SAE 30 oil for an application, hot ambient temperatures might indicate the use of an oil with a:
   a. Higher viscosity
   b. Lower viscosity
4. If you increase the speed of a bearing, would you tend to recommend an oil that is:
   a. Higher in viscosity
   b. Lower in viscosity
   c. The same viscosity

5. If you increase the ambient temperature of the bearing, would you tend to recommend an oil that is:
   a. Higher in viscosity
   b. Lower in viscosity
   c. The same viscosity
6. If you increased both the speed and the temperature of a bearing, would you tend to recommend an oil that is:
   a. Higher in viscosity
   b. Lower in viscosity
   c. The same viscosity
7. Metal fines in the oil do not contribute to oil oxidation.
   a. True
   b. False
8. Foaming in a circulating oil system is an indication of a defective product.
   a. True
   b. False

3
Application of Lubricants

Bearings, Cylinders, and Gears

The application of lubricants is very simple. There are only three things that ever need lubrication. They are (1) bearings, (2) cylinders, and (3) gears.

"Bearings" include antifriction bearings, plain bearings, and even bearing surfaces like the ball joints of a car or truck.

"Cylinders" include the cylinders in an engine, compressor, or hydraulic system.

"Gears" include a gear set which transmits power or motion such as a transmission, differential, or worm gear reduction.

In the final analysis, these are the only components which need lubrication. How these components are designed and used and the conditions which exist in and around the application will dictate the type of lubricant and the method of application. We return again to the right lubricant in the right place at the right time to provide proper lubrication.

Bearings

Oil-lubricated bearings may be designed (1) to use an oil cup for once-through application, (2) to have an oil reservoir so oil is picked up and distributed by the bearing or some device, or (3) with a pressure oil circulating system. The pressure oil circulating system will provide the maximum protection and lubrication for an oil-lubricated bearing.

There is no standard oil viscosity recommendation. The recommendation varies for each type of bearing or application depending on material, size, speed, and temperatures. The ISO grades 46, 68, or 100 oils with rust and oxidation inhibitors are close for normal speeds and temperatures. When abnormal conditions exist, use the following guidelines.

For *lower* speeds, use a heavier oil.

For *higher* speeds, use a lighter oil.

For *lower* temperatures, use a lighter oil.

For *higher* temperatures, use a heavier oil.

These changes from the "normal" oil will provide that bearing with the correct operating viscosity under those different operating conditions.

When a bearing is grease-lubricated, the design may permit regreasing, be hand packed, or be a sealed or "packed for life" bearing. Regreasing may be accomplished with a grease gun or by hand packing the bearing. The packed for life bearing has a larger than normal grease cavity and is packed for the life of the lubricant rather than for the life of the bearing. Sealed bearings are very good for very dirty or corrosive environments because not only do they seal the lubricant in, they seal dirt and corrosive fumes out.

If a bearing is removed and hand packed, care should be taken to apply only enough grease to fill approximately 50 percent of the space between the balls or rollers. New grease should then be applied to the bearing housing so that the grease supply is above the bottom of the bearing. This will allow for plastic flow of the grease at operating temperature.

If a bearing has a zerk fitting in it, it can be regreased with a grease gun. In the bottom of that bearing housing, a purge fitting or grease plug may be found. This plug should be removed when grease is applied to the zerk fitting so the old contaminated grease may be purged out the bottom of the bearing. After the new grease has been applied, the purge plug should be left out for a while during operation to allow any excess grease to heat and escape.

Excess-pressure grease fittings are also available and should be used, particularly if a bottom relief plug is not present. These fittings come in various psi ratings and are designed to prevent excessive pressure from building up in the bearing and blowing the seals out.

A ball bearing prefers a stiffer grease than a roller bearing. As the ball bearing begins to rotate, the grease moves to the outside of the bearing race. Oil then bleeds from the grease to lubricate the bearing surfaces. The grease should stay on the outside of the race so the balls do not have to plow through grease if it were to slump back into the bearing.

Roller bearings, on the other hand, seem to prefer a small amount of a softer grease to slump back into the bearing and then be thrown back to the outside of the race. This slumping action improves the lubrication of the rollers.

Remember, the grease itself does not lubricate. The oil bleeding from the grease does the lubrication.

More greased antifriction bearings fail from overgreasing than from any other cause! *Do not overgrease any antifriction bearing.* A grease cavity 50 percent full is the maximum amount of grease which should be in an antifriction bearing. For more information on grease types and the oil in the grease, refer to Chapter 9.

Cylinders

Cylinder lubrication is often ignored because most cylinder lubrication is secondary to the lubrication of the bearings in engines and some compressors. Many cylinders are lubricated by the splash of the lubricant from the crankshaft. In some large compressors, very small amounts of oil are injected into the cylinder area to ensure the proper lubrication of the cylinder walls. The lubricant in the crankcase is normally also used for the cylinders. Makers of some large air compressors recommend a steam cylinder oil with fat be used for proper break-in prior to using the recommended lubricant.

A very thin oil film is adequate to protect the cylinder walls. Do not overlubricate cylinders. Two single cigarette papers can be used to wipe the oil from a cylinder wall. Upon withdrawal, only one of the two papers should be oil-wet. If two are wet, too much oil is probably being injected into the cylinders.

In some large compressors, which are handling wet or sour gas, the special cylinder lubricant mentioned above is used. The fatty compound in the viscous steam cylinder oil helps prevent the oil film on the cylinder wall from being penetrated or affected by the moisture or corrosive elements.

Gears

Gears transfer power and/or motion from the power source to the application. Gears can be used to change the direction, speed, or torque of the motion. While most gears are enclosed in a gearbox, some are large open exposed gears.

The gear teeth must be separated by a lubricant as the teeth come into mesh. In some heavily loaded gear applications, the lubricant must contain extreme-pressure additives to protect the gear teeth. Without the additives, the viscosity of the lubricant alone is not enough to prevent metal-to-metal contact and excessive wear.

In gear cases which contain finely finished precision gears, are well engineered, and are not subject to extreme pressures or shock loading, a straight rust- and oxidation-inhibited turbine-quality gear oil is recommended. Since this type of oil has a better oxidation life than do the extreme-pressure oils, gear manufacturers make that oil their preference. They go on to state, however, that should the application involve shock loading, an extreme-pressure oil is to be used.

Unless customers have a number of this type of gear sets, they will normally consolidate their requirements into extreme-pressure gear oils of several different viscosities.

In worm gear applications, there is more sliding motion than the rolling motion found in other gear sets. A more viscous lubricant is needed to separate the gears. A special fatty compound additive is added to the viscous oil to protect the sliding gears from metal-to-metal contact. Compounded steam cylinder oils like those used in some compressor cylinders are normally used in this type of gearing for added lubricity. Extreme-pressure additive will not protect worm gears as well as a fatty compound because of the sliding motion involved. The active extreme-pressure additives could corrode the nonferrous parts at elevated temperatures.

The use of synthetic-based gear oils is gaining favor with worm gear manufacturers because of the reduced internal fluid friction of these fluids. Manufacturers are able to uprate their gearboxes and fill them for life with synthetic gear oil.

In some heavy industry, an NLGI 000 semifluid grease is being used in leaky boxes until they can be overhauled. The semifluid grease clings like honey to the gear teeth and provides good lubrication while minimizing the leakage. It is practically impossible to change the lubricant, however, because of the grease-type structure. It makes the removal of normal metal fines very difficult and should be considered only as a stopgap measure or in very rough coarse gearing.

Open gears are lubricated with an asphaltic- or grease-type compound, with a very heavy oil, which is thick and tough enough to stick on the gear teeth and separate them as they go through mesh. These compounds may be diluted with a solvent or heated for easy application and contain extreme-pressure additives and/or fillers such as moly or graphite. Lubrication of open gears is, at best, very difficult to do efficiently.

Inefficiency in a gear set produces heat. The temperature rise in most gear sets is about 50°F over ambient temperature. The temperature rise in a worm gear set is about 90°F over the ambient temperature because it is less efficient and generates more heat than other types of gear sets. This temperature rise can be reduced with synthetic fluids. The manufacturer takes this factor into consideration when lubricant recommendations are made.

Operating Conditions

Many customer locations are ideal operating conditions. Lubrication inside a nice clean plant with normal room temperatures and regular maintenance is ideal. Although these conditions are common in many industries, they are very uncommon in others, e.g., steel mills, foundries, open-pit mines, and construction sites.

We often encounter applications that are very hot—like a gearbox next to a furnace where the ambient temperature is 120 or 130°F and the gear set lubricant is 170°F and higher. The oil in the gear set will oxidize rapidly because of the high temperature, and the customer will have to change the oil more frequently or switch to a synthetic gear oil. If the shortened drain interval is ignored, the gear set will soon fail and the inside of the box will appear as if asphalt was used as a lubricant. Your oil supplier will be able to assist you with unusual applications.

Cold temperature can be a problem with the start-up of machinery. Accelerated wear will occur at start-up if an incorrect lubricant is used. Multiviscosity engine oils, low-pour or multiviscosity hydraulic fluids, and softer greases should be used in colder climates. Synthetics could also be considered for additional protection. Watch carefully for condensation in cold weather applications. Shortened drain intervals may be in order to protect the equipment. Extreme-pressure additives normally require heat to activate, so special additives may be necessary in cold climates.

Dirt is a major enemy for equipment. If dirt enters a gear set, it will combine with the oil to form a grinding compound and cause rapid wear of the gear teeth and bearings. Shortened drain intervals can reduce this wear.

A far better solution is to prevent the entry of the dirt into the gear set. Check the breather and the oil plug to ensure they are tight and sealed.

Check the oil storage and the container the oil was in prior to adding it to a gear set, hydraulic system, engine, or circulating oil system.

The injection of dirt is also a problem with a grease application. A dirty zerk fitting will inject dirt directly into the bearing if it is not cleaned prior to applying the grease. Check the main container of grease or oil for exposure to dirt while in storage. The container should never be left open.

Troubleshooting

Dirt, extremes of temperature, extended drain intervals, and poor handling practices are the main sources of lubricant contamination and short component life. Several other potential problem areas should also be considered.

1. Wear debris from a failure in a system must be cleaned out prior to refilling the system with lubricant. If it is not cleaned out or flushed, the wear debris will quickly cause another failure.

2. Observe how the oilers handle the oil for various applications. Many times the wrong oil will be put in an application because it was the "closest" oil.

3. If you have a product performance problem, check if the product is stored in such a way as to protect it from water and dirt. Also check the container used to deliver the oil to the application and check if oil coolers or oil heaters are involved in the application. Look for that outside problem source before you look at the oil.

4. When you take an oil sample, give it the "eyeball" test before you send it to a laboratory. If the oil looks milky or water is evident, why send it to a lab to tell you that fact. If you have foam, ask if a defoamant has been added. If it has, too much could have been added and this would cause "air entrainment," which is a thick layer of very small bubbles. The only solution to this problem is to drain and refill the system. Check for an air leak at the suction side of the pump as a major source for the air entering a circulating system and causing the foam.

5. When you are notified of an oil-related problem, do not be too quick to blame the oil or grease. It is suggested that your maintenance team investigate all possible avenues of contamination and adverse operating conditions. There could easily be outside factors which have contributed to the problem.

6. When taking an oil or grease sample, use a clean container. If the oil is a light-colored oil or if water or coolant is suspected, use a clear glass container. It may be possible that the contamination can be viewed through the container. You will also be able to inspect for dirt or grinding dust as it settles to the bottom of the container. You will be able to uncover the causes of many oil-related problems and effect a solution before you could obtain outside assistance.

Problem solving should involve the following:

1. Gather information about maintenance procedures, product handling, and storage and confirm that scheduled maintenance has been performed.

2. Get at least one sample of the product from the system in question and one from that source container. Inspect the samples for contamination.

3. If conditions in and around the problem application do not pinpoint the cause, send both a new product and used product sample to a laboratory for analysis. There might be something wrong with the new oil you are putting in the application.

Self-Quiz

Mark your answers to the questions and compare them with the answers listed at the end of the book.

1. In bearing applications, select the most nearly correct answer.
   a. Higher speeds require a heavier oil.
   b. Higher temperatures require a lighter oil.
   c. Lower speeds require a lighter oil.
   d. Lower temperatures require a lighter oil.
2. What is a major cause of antifriction bearing failure?
   a. Fretting corrosion
   b. Lack of maintenance
   c. Overgreasing
3. Ball bearings prefer a grease which is
   a. Stiffer than a roller bearing
   b. Softer than a roller bearing
4. In a compressor cylinder handling wet or sour gas, what additive would be used to protect the cylinder walls?
   a. Extreme pressure

   b. Fatty compounds
   c. Detergents
5. What temperature increase over ambient would you expect to see in a worm gear set?
   a. 50°F
   b. 70°F
   c. 90°F
6. Most wear will occur during
   a. Cold starts
   b. High-temperature operation
7. What do you really want your oil supplier to provide you?
   a. Grease
   b. Diesel fuel
   c. Oil
   d. Lubrication

4
Lubricant Formulations

Crude Oil

Crude oil refined in the United States comes from many different sources, including the states of Texas, Oklahoma, California, Louisiana, Alaska, and several others as well as offshore. Much of our crude also comes from Mexico, Canada, South America, and the Middle East. With a daily consumption of approximately 17 to 18 million barrels (42 gallons per barrel), any interruption in one source can be very inconvenient for the oil and other industries.

The United States woke up to that fact when OPEC (Organization of Petroleum Exporting Countries) embargoed the United States in the early 1970s. It should be no big surprise when and if it happens again in the 1990s. As the largest industrialized nation in the world, our whole economy hangs on the whims of these third world countries. This is a very unstable situation for our economic well-being and will continue to be in the future.

Crude oils are found in a variety of types ranging from light-colored grades which yield primarily gasoline to black crudes which are nearly solid asphalt. The hydrogen carbon atom structures vary a great deal as do the impurities, such as sulfur or wax, in the crude oils. Some crudes are suitable only for the manufacturing of gasoline, diesel, fuel oil, and by-products while others are preferred for the manufacturing of fine lube oil stocks.

There are three basic types of crudes:

1. Paraffinic
2. Intermediate, naphthenic, or mixed-base
3. Asphaltic

The different basic crude stocks may be more suitable for one type of application than for another. It would not be cost-effective to use a fine paraffinic stock oil for open gear lubricants where a poorer-quality asphaltic stock oil would perform fine. Likewise, to use an asphaltic stock oil for high-performance hydraulic oils would be foolhardy.

Lube-Oil Stocks

Paraffinic and intermediate (mixed) stocks tend to make up the majority of the lube oil stocks today. Even then, the stocks are selected and segregated according to the principal types of hydrocarbons present in them. The types of lube oil stocks present, after distillation, are carefully monitored, tested, and selected for certain applications, such as motor oil, hydraulic oil, turbine oil, and grease applications.

A lube oil stock from paraffinic crude would contain wax and have a higher pour point because of the wax crystals, even though much of the wax is removed as part of the refining process. Paraffinic stock oils also have a higher viscosity index, which makes them more suitable for applications such as hydraulic and automotive products which are used over a wide temperature range and require good natural viscosity index and oxidation resistance.

A lube oil stock from naphthenic crude (mixed) contains essentially no wax and therefore will have a much lower natural pour point. It will also have a very low viscosity index, making it less suitable for a lubricant which must operate over a wide range of temperature. Naphthenic stock oils are typically used in products such as some gas engine oils, refrigeration oils, and railroad engine oils.

The continuing availability of one base stock versus another is uncertain, and suitable alternative stock oils are always being sought by the oil companies.

The different types of crude oil allow refiners to select those which, when fully refined, will provide base stocks suitable for their needs. Quality and types of stock oils vary, and they are traded and purchased between major companies according to their needs.

The refining process is much too complex to detail at this time. Many processes are used to remove most of the undesirable elements from the stock oils to make them more suitable as a lubricant. These refining processes can produce by-products, such as wax, which can be profitably sold to industry. Millions of pounds of wax are consumed yearly by the companies who make paper matches. A stock oil may also be subject to solvent extraction and/or hydrofinishing to produce a finished base stock oil.

In addition to chemical and physical characteristics, a base stock oil must be compatible with the additives which are commonly used in lubricating oils for particular functions.

Companies use different-quality base stocks to manufacture their finished products. Some companies might use a very high quality base stock with some additives to produce a product while another might use a medium-quality base stock with more additives to make an "equivalent" product. This is one of the reasons there may be differences in the performance and the price of an "equivalent" product among competitors.

There is an old saying which is true of many petroleum products and so many other things:

> IN MOST CASES, YOU GET WHAT YOU PAY FOR.

A more highly developed type of stock oil is just starting to appear in the marketplace today. It is a superhydrotreated paraffinic stock oil that is relatively free from foreign and undesirable elements, has a pour point below 0°F and costs more than conventional lubricant stocks but less than the synthetics. It could be an ideal stock to use in products used under severe conditions, and with the proper additive package, such as silicones, feed rates of the lubricant can be reduced in some applications to make it very cost-effective. Time will prove it to be a very good middle-ground product in motor, hydraulic, and compressor lubricants.

Common Additives

Additives are used to enhance the performance of a petroleum lubricant. In most cases, the enhanced performance in one area may be slightly offset by a decrease in performance in some other area; e.g., adding an extreme-pressure additive to a gear oil will increase the extreme-pressure capabilities but slightly decrease the oxidation life of the product.

The addition of some additives may also reduce the demulsibility, which would be undesirable in an application where water contamination is normal and water separation is important. Should this be a potential problem in a circulating oil system, it should be well noted by the oil supplier and equipment manufacturer.

Some additives are a "give-and-take" proposition. You gain a little here and give up a little there. The key is that the gain should be where it is needed and the loss should be in an area where it is not important or can be controlled in the application.

Additives have allowed lubricants to perform better and longer in severe applications. The advanced technology in today's machinery and internal-combustion engines requires the additives common in premium lubricants to function properly. Practically all lubricants contain at least one additive and most contain several different types. Some additives impart new and useful properties to the oil, like rust protection; some enhance properties already present, such as antiwear; while others retard undesirable changes which take place in service, such as oxidation inhibitors.

The effectiveness of additives reaches a point of diminishing return. To add twice the additive will not give you twice the results. The addition of an aftermarket can of additives can improve a poor-quality product but will be a waste of money in a quality product. The exception to this would be a *defoamant* which reduces the surface tension of the fluid and allows the air to escape easily from the fluid. Adding too much defoamant will cause air entrainment, the entrapment of tiny air bubbles in the fluid, which will destroy the fluid for further service.

Some of the most common additives for lubricants follow.

Oxidation Inhibitors

These types of additives decrease the rate of oil oxidation by reducing the tendency of the hydrocarbons to combine with oxygen molecules, and as a secondary effect, they can reduce the tendency to corrode certain types of sensitive bearing materials.

Oxidation increases a lubricant's viscosity and causes a chemical change which results in the formation of peroxides and organic acids. These chemical changes produce elements which cause corrosive action. High temperatures, catalytic metals, and exposure to air increases the oil's rate of oxidation. After the oxidation inhibitors are depleted, the rate of oxidation depends on the natural oxidation resistance of the base oil. A quality base oil will then oxidize more slowly than will a poorer-quality base oil.

Some oxidation inhibitors are heat-sensitive and are rapidly depleted with higher than normal temperatures. In those cases, a different type of oxidation inhibitor may be used to extend the useful life of the lubricant.

Pour-Point Depressants

Oils containing paraffinic hydrocarbons tend to form wax crystals at moderately low temperatures. These wax crystals prevent the oil from flowing. This is undesirable in applications where start-up may be at temperatures lower than the pour point of the oil.

Although many of the paraffin elements are removed by dewaxing in the refining process, some remain in the oil. Wax, as a by-product, has become a salable and profitable by-product for the oil companies.

Pour-point depressants are high-molecular-weight polymers which inhibit the formation of the wax crystals and therefore allow the oil to flow at a lower temperature. They will not entirely prevent the formation but will lower the temperature at which the wax structures start forming.

Rust Inhibitors

These additives are used is very small amounts, and the type used varies with the finished product's application. Some rust inhibitors improve the oil's ability to adhere tenaciously to steel surfaces and therefore prevent moisture from penetrating the protective oil film.

These are typically used in bearing lubrication, steam turbine shafting, gears, and machine tool spindles. Others may act as emulsifiers which will combine with the moisture and prevent it from making contact with the metallic surface. This type might be objectionable in applications where demulsibility of the oil is desirable, such as rolling oils, or other applications where water contamination can be a problem. The use of rust inhibitors is very common.

Antiwear Additives

When machine components operate under boundary lubrication (where metal-to-metal contact can and normally does occur) antiwear additives are necessary. Good examples are high-pressure hydraulic systems and engines. Zinc dithiophosphate or zinc dialkyldithiophosphate are commonly used antiwear or antiscuff additives which combine the effect of an antioxidant and corrosion inhibitor with antiwear protection. They are classified as a mild extreme-pressure or friction-reducing additive. Tricresyl phosphate has also been used in some hydraulic and circulating systems.

Extreme-Pressure Additives

Where film strength and extreme-pressure characteristics are needed to combat high or shock loading of components, sulfur-, phosphorus-, and/or chlorine-based additives are used. A sulfur-phosphorus combination is the most commonly used extreme-pressure additive especially in automotive and industrial gear oils.

Extreme-pressure additives are activated by heat rather than pressure. They chemically react with metal surfaces to form a film, or surface compound, which gently wears or polishes off rather than allowing the mating surfaces to weld and tear loose, causing destruction. It takes time for this compound to develop in service. After it has developed, it has a lower shear strength than the base metal, thus reducing friction. The additive compounds, activated by heat, could tend to be corrosive to nonferrous parts and/or act as a prooxidant and somewhat shorten the oxidation life of the product. Special low-temperature extreme-pressure additives are necessary in very cold applications.

Detergents and Dispersants

These are the additives commonly used in engine oils to disperse the particles of contamination created by combustion and help prevent the formation of harmful deposits, such as sludge and varnish. Regular drain intervals and filter changes are essential to the prevention of the particles depositing in the engine. The filter removes many of these particles and products of oxidation which would tend to accelerate the rate of oxidation. Overextended drain intervals or excessive by-products of combustion can cause the depletion of these additives and allow the contaminants to deposit themselves throughout the engine. Detergent-dispersant additives are used more to keep an engine clean than to clean up deposits already in the engine.

Calcium, magnesium, or barium soaps of petroleum sulfonic acids or synthetic sulfonic acids are some of the materials which may be used. Various compounds of barium, sulfur, and phosphorus-containing polymeric and ashless detergents are also used. They are commonly found in motor oils and, to a lesser degree, some hydraulic oils.

The addition of ash-type detergent additives will increase the ash residue and can be detrimental in applications where increased ash can cause increased maintenance, such as an air compressor. In engine applications, where acidic conditions can easily occur, ash-type additives provide a higher and more stable TBN (total base number) to more effectively combat corrosive wear of the engine components.

Ashless-type detergents and dispersants are very effective and are used in many modern engine oils today, but some have found that when they are used alone, without the ash type also included, some rust protection is lost. This is of little concern for engines which are operated continuously or daily.

Viscosity Index Improvers

An oil thins when it is heated and thickens when cooled. The resistance of an oil's viscosity to change with temperature is called the "viscosity index." The higher the viscosity index, the more resistant an oil is to thinning with increased temperature.

A viscosity index improver will increase the oil's resistance to thinning with increasing temperature. Thus, a *high* VI oil will resist thinning better than a *low* VI oil. An SAE 10W oil and an SAE 40, with the same viscosity index, will thin at the same rate with increasing temperature. If a viscosity index improver is added to an SAE 10W oil, increasing its resistance to thinning, it could be as thick or viscous as an SAE 40 weight at an elevated temperature. That oil could then be called an SAE 10W-40. The resulting oil would be like an SAE 10W at low temperatures and like an SAE 40 at elevated temperatures. It has resisted thinning with heat and therefore now has a high viscosity index.

VI improvers are used in some hydraulic oils, transmission fluids, motor oils, gear oils, and many other applications. Most mineral oils with a viscosity index over 100 contain a "viscosity index improver."

The viscosity index improvers of today are far superior to those of the 1950s or 1960s. Some early multiviscosity 10W-40 motor oil experienced VI improver shearing which would quickly make that 10W-40 a 10W-30 or even lower with use. The ability to keep that 10W-40 a 40 at elevated temperatures has greatly improved since that time.

Defoamants

An oil's tendency to foam will depend on the type of crude, refining, viscosity, and application. If the designated application has a tendency to aerate the oil, a very small amount of defoamant will be added in the formulation of the product. A defoamant will reduce the surface tension of the oil, allowing the small air bubbles to combine into large bubbles and then dissipate into the air. Every oil will foam, and only the oil's lack of ability to expel the entrained air would classify it as "foaming."

The addition of too much defoamant has been described above. Only the proper amount of defoamant should be added to an oil system, never more than that.

Tackiness or Stringy Agents

These types of additive are added to promote adhesion between the lubricant and the point of application. They are normally added to "nondrip" oils that would be applied to an overhead chain drive or to a grease to impart that "tacky" property. If installed in an application which subjects the lubricant to a shearing action, the tacky property of the lubricant will soon be sheared away or destroyed.

Fatty Oil Agents

These additives are commonly used in applications where water might tend to displace the oil from the lubricated surface, as in a steam cylinder or some large compressor cylinders. They also impart additional "slipperiness" or "oiliness" to the lubricant to provide better lubrication where the application involves a sliding action, as in many worm gear sets. The original source for this additive was the sperm whale but that source is obsolete and other forms, such as rapeseed oil, oxidized wax products, and tallow, are now used.

Grease

Grease is thickened oil which contains many of the same additives as do fluid lubricants. Theoretically, grease does not lubricate. The oil bleeding from the grease does the lubrication. Additives will enhance the performance of the oil in many applications.

Greases contain these elements:

1. An oil chosen to perform a particular function or chosen to be multi-

purpose in multiple applications. A grease designed to be used in a rough hot application should use a viscous oil, and a grease designed for high-speed bearings would use a far less viscous oil. Fluids other than mineral oils may be used as the lubricating medium in specialized applications.

2. A thickener which has the properties desired in that grease. High-temperature resistance, waterproofness, or excellent mechanical stability are examples. The amount of thickener used will normally determine the NLGI grade or penetration.

3. Additives to enhance the performance of the oil under adverse operating conditions. These might include moly, graphite, paratac, extreme-pressure additives, and rust and oxidation inhibitors. There are other additives or fillers, such as dyes, which do nothing but make the grease appear unique and different for marketing purposes.

In most cases, grease is used in applications where oil will not stay. Grease around seals prevents dirt from entering better than oil. Grease, however, does not cool as efficiently as an oil and cannot be circulated. It is applied and remains in the application until (1) more is applied and the old grease is flushed out or (2) the bearing is removed, cleaned, relubricated, and put back into service. A bearing, sealed for life, has a larger grease cavity and is sealed for the life of the lubricant in it.

Self-Quiz

Mark your answers to the questions and compare them with the answers listed at the end of the book.

1. A naphthenic crude contains wax which increases its pour point.
   a. True
   b. False
2. An extreme-pressure additive is the same as an antiwear additive, but more of it is used in the finished product to provide added protection.
   a. True
   b. False
3. Extreme-pressure additives perform best at low temperatures.
   a. True
   b. False
4. The additive commonly used in an oil to reduce its tendency to thin with heat is called:

  a. Detergent and dispersant
  b. Viscosity index improver
  c. Antiwear

5. The basic properties of a grease are dictated by what "part" of the grease?
  a. Thickener
  b. Viscosity of the oil
  c. Additives used

6. A viscous oil is required to make a grease.
  a. True
  b. False

7. Viscosity index improvers are used only in multiviscosity engine oils.
  a. True
  b. False

5
Engine Oils

Society of Automotive Engineers

The Society of Automotive Engineers (SAE) has established viscosity classifications for engine oils. These classifications are in the Appendix of this book.

Viscosity grades denoted as W, like 5W and 10W, are measured at low temperatures with a minimum viscosity requirement at 100°C. Viscosity grades without the W, like 30 and 40, have minimum-maximum viscosities, which are measured only at 100°C.

Multiviscosity Engine Oils

If an engine oil meets the SAE 10W specification at a lower temperature and SAE 30 at 100°C, the oil has an SAE grade of 10W-30. The same is true with 5W-30, 10W-40, 15W-40, and the other multigrades. You may see an oil which might be an SAE 20W-20 which could be correctly graded. Mineral-based oils, which meet the viscosity requirements for a W grade and a non-W grade, are classified as "multigrade" or "multiviscosity" and usually contain a viscosity index improver. A 20W-20 and some 10W-20 oils would be exceptions to this rule.

In the section under additives, viscosity index improvers were discussed. VI improvers are added to an oil to increase its resistance to viscosity change with temperature. Some oil companies have used VI improvers that have not been "shear-stable." This means the VI improver would shear, or be cut apart, in service. It can be pictured very easily by visualizing a bunch of "pigs' tails" in the oil. As the oil, with these "pigs' tails," circulates through the engine, they can tend to be cut into small and less efficient pieces. When this happens, a 10W-40 becomes a 10W-30 or 10W-20. The shearing can also affect the lower end, and since we had to start the formulation with an oil lighter than SAE 10 and because the VI improver also thickens the oil, we might even see a 5W-20.

Many manufacturers of quality multiviscosity conventional engine oils now sonically shear their VI improver before adding it to the oil. This, along with today's improved VI improvers, eliminates the initial shearing of the additive and basically reduces or eliminates the problem in the finished product.

American Petroleum Institute

The American Petroleum Institute (API) is an oil industry group that has developed, in conjunction with the American Society for Testing and Materials (ASTM), oil designations which denote the specification or group of specifications and performance levels an oil should meet. They are really the consumer's guide to the expected level of performance of an engine oil.

The two classifications are S for service station, gasoline, or spark ignition and C for commercial, diesel, or compression ignition. The various S and C classifications are listed in the Appendix of this book. The most common API service classifications today are SG for gasoline (some SF oils are still made) and CD, CD-II, and CE for diesel engines. The latest diesel classification, effective Dec. 25, 1990, is currently CF-4, which exceeds the CE requirements and will replace CE in some cases.

Gasoline Engines

Many oils satisfy the API classification of SG, so it is not a problem to pick one. The thing left to choice is an SAE viscosity grade to satisfy the various engine requirements and the most up-to-date specifications. Old habits and preferences are very hard to change, and any change in type or brand name will meet with resistance.

The API rating of SG became tougher in 1993 as the API tightened their formulations involving different base stock oils in the manufacturing process. This is one reason so many companies came out with "new" formulations and synthetic engine oils during 1992.

Multiviscosities like 10W-30 and 10W-40 continue to dominate the gasoline engine oil market despite the fact that many car manufacturers are now recommending 5W-30. Old ways of thinking die very hard. The old recommendation of 10W-40 has given way to lighter viscosities because there were some adverse effects of using too much viscosity index improver in some oils. Despite the changes in gasoline engines over the past several years and the reduction in manufacturer's viscosity recommendations, many old die-hards want that heavier-viscosity oil. If they could buy an SAE 60 weight, they would use it. Some operators of heavy equipment continue to stay with monograde engine oils in their gasoline engines to eliminate the stocking of two grades of engine oil.

There is some evidence that when the diesel engine oil additive packages are changed to meet the more stringent diesel requirements of CE, performance in gasoline engines, even with those oils which are rated SG, will be reduced. This is the reason many of the higher-rated oils for diesel engines are rated only SF rather than SG or the next gasoline engine rating. Ratings such as SA, SB, SC, SD, and SE are obsolete.

Diesel Engines

More and more diesel engine manufacturers are mandating the classification and the viscosity of oil for their engines. The most common viscosities used in diesel trucks are SAE 30, 40, and 15W-40. Engine manufacturers such as Caterpillar, Cummins, Detroit Diesel (four-cycle engines), and Mack prefer SAE 15W-40. These manufacturers also state that single viscosities are acceptable within certain temperature limitations. Exceptions to the multiviscosity preference are Detroit Diesel (two-cycle engines) and Navistar (formerly International) who prefer single-viscosity oils but will accept multiviscosity in colder temperatures. In the large Detroit Diesel 149 engines, an SAE 40 or SAE 50 CE oil is recommended for heavy-duty service under certain temperature restrictions. The higher ash level in this type of oil can and has caused excessive deposits in two-cycle engines.

The present API diesel classifications are CC, CD, CD-II, and CE. CD-II is CD rated plus passing the Detroit Diesel 6V53T test. CE is CD rated plus passing the Mack EO-K2 and the Cummins NTC 400 tests. A CE oil may or may not be rated CD-II. As stated, the newest diesel classification, at this time, is CF-4, which exceeds the service requirements of CE.

General Recommendations

Needless to say, Detroit would tend to prefer CD-II for both their two- and four-cycle engines while Mack and Cummins would tend to prefer a CE or CF-4 rated oil. Each will recommend the oil which meets the specifications of their particular engine. A diesel engine oil rated CD II and CE would satisfy both and would be an excellent choice for a mixed engine fleet. These preferences are always subject to change and undoubtedly will change as time goes on.

The customer is going to prefer one type of oil over another. Temperature considerations will also be a factor in the choice of viscosities. Truckers will lean toward the 15W-40 engine oils because their most common engines recommend it and their engines will typically be exposed to a variety of operating temperatures. Contracting and mining accounts will tend to lean toward the single viscosities like SAE 30 and 40 in reasonable climates with an SAE 10W or 20W-20 for hydraulic and steering fluid. In colder climates, 15W-40 could be considered as the better engine oil based on startability and engine protection. In hot climates, a mine should consider the use of an SAE 40 or 50 for heavy diesel equipment and consolidate with an SAE 30 for gasoline engines and mobile hydraulics in heavy equipment.

Be aware of the sulfur content of the diesel fuel in your area. Some parts of the country have higher content than others. In some parts of the West Coast the sulfur content of diesel is less than 0.1 to 0.2 percent, and other parts of the country may have a sulfur content of 0.5 percent or more. Beware if diesel fuel is being shipped out of Mexico because the sulfur content can be well over 1.5 percent. Normal drain intervals and low TBN oils will dramatically increase engine wear with this type of high-sulfur diesel fuel.

The higher the sulfur content of the diesel fuel, the more corrosive the combustion by-products can be to an engine. The answer to living with that condition is shortened oil drain intervals and higher TBN engine oils.

Many major oil companies are touting their high TBN engine oils as the savior of diesel engines. Their claims are valid in many cases and areas, yet an overkill in other cases where short drains, dirty conditions, and low-sulfur diesel fuel are normal. Weigh the operating conditions and sulfur content of the diesel fuel when making your choice of engine oils.

New API Classifications for Diesels

The newest proposed API diesel ratings are CF, CF-2, and CF-4. CF would probably pass the new Cat 1-K test plus the L-38 test. CF-2 will be CF plus the newer Detroit Diesel 6V92TA test. CF-4 will be CF plus the Mack EO-K2 test and the new tighter Cummins NTC 400 test. Although CF-4 was approved in December 1990, it is now doubtful the CF and CF-2 ratings will be accepted. This rating game will continue as long as the engine manufacturers continue to get more performance out of their engines and are expected to also reduce emissions. The new engines will gain more and more in performance, and the engine oils must keep up to protect them.

MIL Specifications

In the 1960s and 1970s military specifications were a common method of determining the service level of diesel engine oils. In the 1980s and on into the 1990s, engine manufacturer specifications and engine tests have dominated and will continue to do so.

Many of the old MIL specifications are obsolete, but the most current "obsolete" ones are MIL-L-2104B, C, and D, replaced by MIL-L-2104E, and MIL-L-46152C replaced by MIL-L-46152D.

MIL-L-46152D is for both gasoline and diesel engines in commercial vehicles used in the federal and military vehicles. The gasoline performance would be at the API SG level while the diesel engine performance might only be at MIL-L-2104B or API CC level. Some of these oils will also be rated SG/CD, CD II, and CE.

MIL-L-2104E is for both gasoline and diesel engines in military tactical vehicles. This oil would also meet the Cat TO-2, Allison C-3 specifications and API service classifications of SG/CD II and CE. Some will also meet SF/CD, CD II, or any such classification.

Today, readily available engine oils from the major oil companies easily meet these specifications. These same oils, however, may not be "qualified" (certified by the military) against the specifications but are formulated so that they would meet the specification. If you are dealing with military customers who demand that oils are certified against a particular MIL specifications be very sure you understand exactly what they mean and what is required so you can inform your supplier.

CCMC, CLCA, and ACEA
Organizations

Up until the end of 1990, there was an organization of European automotive manufacturers called the CCMC (Comité des Constructeurs des automobiles du Marché Commun). This organization consisted of twelve European car markers who all had individual veto power of any proposals presented to the group. Eleven of the twelve members resigned in protest of one member's vetoing a proposal all the other members considered important. This group did not include Ford of Europe or General Motors of Europe. Their purpose was to present a common front against Japanese imports and establish standards in their industry.

Work started immediately to establish the CLCA (Comité de Liaison de la Construction Automobile) with few differences from the CCMC except for the elimination of the veto and adoption of an 80 percent majority rule. This organization was short-lived and was replaced by the ACEA (Association des Constructeurs European Automobiles) with a different structure and membership which included Ford and General Motors. The ACEA will be headquartered in Brussels.

It is expected the ACEA will reissue some of the gasoline and diesel oil specifications started by the CCMC group which currently include G-4, G-5 for gasoline engines and PD-2 for diesel and D-4 and D-5 for industrial engines.

The CCMC oil specifications had little impact in the American marketplace. The specifications issued by the new ACEA will, however, become an important factor in engine oils in the United States because Ford and General Motors are now included as members in that group.

Synthetic Engine Oils

Synthetic and semisynthetic engine oils are now receiving a lot of attention. There are advantages and disadvantages to these types of products depending partly on their use.

Semisynthetic oils are a combination of a synthetic fluid and mineral oil. There will be some reduction in friction both in the parts being lubricated and within the fluid itself. The natural film strength of the combination should increase over straight mineral oil. Claims of higher oxidation resistance, however, are questionable since the mineral oil portion of the combination will oxidize at the same rate as a comparable mineral oil product. These marginal benefits could be obtained at a cost higher than mineral oil engine oils.

The obvious disadvantages to the semisynthetic are higher cost per gallon, poorer performance than 100 percent synthetic products, and poor cost-effectiveness in applications where dirt is a problem. Many knowledgeable people feel, if you are going to synthetics, go all the way up to 100 percent synthetics and gain all the benefits.

The 100 percent synthetic engine oils are becoming more common today and have properties which can provide superior performance.

Compared to mineral-based engine oils, the oxidation resistance is far superior, the friction in both the application and the fluid itself is less, and the synthetic products have much better low-temperature characteristics. The film strength of the synthetic products is also better than that of mineral oil engine oils. The natural detergency of the base synthetic fluid plus the detergent-dispersant additives will clean up an engine and keep it cleaner during service.

The synthetics have equal or superior characteristics compared with mineral oil-based engine oils. They cost more but in most cases can be very cost-effective.

If you carefully track fuel economy, it is possible to notice up to a 5 to 10 percent improvement in both gasoline and diesel engines depending on the engine. Oil drain and filter change intervals can be extended unless you are in the warranty period. Even then, an oil analysis program can support you if there is a problem. The total performance in an engine is generally superior to that of mineral-based motor oils. In normal or severe service (trailer towing, etc.) gasoline engines, the synthetics have proved themselves in performance and cost-effectiveness.

Synthetic lubricants can be a great problem solver for automotive fleets when very low or very high operating temperatures are causing failures. In cold climates, engine heater use can be reduced or eliminated because of the superior low-temperature fluidity. In cross-country winter travel, the synthetic will provide easier cold starting and excellent protection in both warmer and colder climates.

The disadvantages to synthetic engine oil are higher cost, roughly 3 to 4 times mineral oil, and poor cost-effectiveness in applications where dirt is a problem. In engines which have higher than normal oil consumption, the synthetics may not be cost-effective because of the high makeup rates. The type of synthetic fluid will determine its effect on common seal materials and its compatibility with mineral oil engine oils. In most cases, there are no problems in this area.

If a synthetic engine oil is used in an engine that has been poorly maintained, the sludge and varnish which formerly helped seal certain areas could be broken loose and seal leaks develop. The owner should stay with a conventional engine oil or repair the engine if it is in that condition.

Troubleshooting

An oil analysis program can provide many clues to problems in engines. If you cannot persuade your oil supplier to include a program in the service, you will find it very beneficial to contract with an independent outside laboratory for this service. The service will quickly show some of these problem areas.

1. Coolant leaks may exist in the engine. The drained oil will have a "coffee with cream" appearance. If the water has been boiled off, only a residue of sodium will show in a lab analysis.

2. Glycol in the engine oil. If there is a coolant leak, the water phase could have evaporated off, leaving the glycol residue behind. That residue will oxidize quickly and a reddish varnish may be present from the oxidized glycol.

3. High soot levels in diesel engine oil are a sign of incomplete combustion and could indicate a plugged air intake system or faulty injectors. Soot levels can be measured by the laboratory analysis. High soot levels are believed to make the engine oil abrasive with resulting high wear in the engine. Oil oxidation and soot increase the viscosity of the oil.

4. Drain intervals may have been extended beyond the manufacturer's recommendation, if desired, with the use of oil analysis. Maybe the oil and filter change were accidentally skipped and were never made. Sludge in the filter and engine would be an indication. Be careful if you see sludge because it also might be from water and/or glycol contamination.

5. The diesel fuel sulfur content might be very high and the ability of the engine oil to handle acidic conditions was exceeded. The bearing material will then appear to be etched. With high-sulfur fuel, high TBN engine oils must be used and drains shortened.

6. Fuel dilution may have reduced the viscosity of the oil to a point where it could not protect the engine. Watch this closely because fuel dilution will thin the oil while soot from combustion will thicken it. If the viscosity is OK but there are high levels of soot, it might mean there could be fuel dilution in that engine.

7. Most progressive fleet companies today use oil analysis. When you are dealing with high dollar items like a fleet of trucks, it is penny wise and pound foolish not to use it. If you are working on a problem, check the past analysis for the starting of a trend as an indication of the problem starting. Trend analysis is much better than spot checks in the determination of engine function.

Self-Quiz

Mark your answers to the questions and compare them with the answers listed at the end of the book.

1. An SAE grade oil followed by W, like SAE 10W, means its viscosity is qualified for "warm" temperatures.
   a. True
   b. False
2. Which multigrade engine oil designation is incorrect?
   a. 10W-50
   b. 10W-20W-30
   c. 5W-20
3. What is the first letter of the API service classification for a diesel engine oil?
   a. S
   b. A
   c. C
4. Whose performance classifications are most common in the marketplace today for a diesel engine oil specification?
   a. Manufacturer's specifications
   b. Military specifications
5. Regular oil analysis is not really needed for a fleet because spot checks with oil analysis can tell you all you need to know about the engine.
   a. True
   b. False
6. The CCMC oil specifications are currently a dominant force in the marketplace.
   a. True
   b. False
7. The most current and updated diesel classification is
   a. CD II
   b. CE
   c. CF-4

6
Automotive Gear Oils

Society of Automotive Engineers

SAE has established viscosity grades or viscosity brackets for automotive gear oil as they have for engine oils. They have established temperature limits, in Celsius, on the viscosity of 150,000 cP (centipoise) for the cold temperature grades (those which are noted W, like 80W). They have also set minimum viscosities at 100°C for the W grades.

For grades which are not followed by a W, they have established minimum-maximum limits on viscosity, measured only at 100°C. The SAE 250 grade, however, has a minimum viscosity only at 100°C. This measuring method is very similar to the SAE grades for engine oil (see the SAE chart in the Appendix). In comparing the engine oil with the gear oil chart, note that, although the SAE 90 gear oil has a higher number than does the SAE 50 engine oil, they have similar viscosities.

The most common SAE grades seen in the marketplace today are 80W-90 and 85W-140. Many operators of heavy equipment use the 85W-140 grade rather than the recommended 80W-90 grade because they feel more comfortable with the thicker oil film. In most cases, this is unwarranted. Both products contain a VI improver (see Chap. 4 under additives) which permits operation across a broad temperature range.

American Petroleum Institute

API has established automotive gear oil service or performance designations for automotive manual transmissions and differentials. The general description of these classifications is as follows:

GL-1. Specified for spiral-bevel and worm gear axles and some manual transmissions under mild service. Usually contain rust and oxidation inhibitors along with defoamants and pour point depressants.

GL-2. Specified for worm gear service more severe than can be satisfied by use of a GL-1.

GL-3. Specified for manual transmissions and spiral-bevel axles under moderately severe service.

GL-4. Specified for hypoid gear service under severe service but without shock loading. This classification is now obsolete. The additive package contained a zinc additive combination, replacing the formerly used lead soap.

GL-5. Specified for hypoid gears under shock loads and severe service. This is the most common grade in use today. It also will meet the MIL specification MIL-L-2105D, in most cases, for a multipurpose automotive gear oil.

Most 80W-90 and 85W-140 gears oils meet the GL-5 classification and the MIL specification without any problem. These products have a high level of extreme-pressure additives and could be mildly corrosive to nonferrous parts in some applications.

GL-6. Although some manufacturers still note it on their containers and product information sheets, this classification is obsolete or was never even adopted by API.

Currently two new proposed API classifications could be effective in the late 1990s.

PG-1. Designed for heavy-duty, high-temperature (up to 300°F) transmissions, with an L-60 cleanliness rating of 9.0+, sludge protection, good seal life, and synchromesh corrosion. PG-1 should become known as *GL-7.*

PG-2. Designed for heavy-duty, high-temperature axles and has the same properties as above but with a cleanliness rating of only 8.5+. The same antiscore as a GL-5 is also required. PG-2 should become known as *GL-8.*

Some of the products presently on the market meet both these requirements with little or no change in formulation. Starting in 1994, most automotive gears oils will be rated as GL-5, GL-7, and GL-8, all in one product. Transmission and final drive manufacturers will then immediately upgrade their recommendations to take advantage of these ratings.

Military Specifications

The current military specification for automotive gear oil, MIL-L-2105D, was issued in 1987. It naturally supersedes 2105B and C, as is the normal procedure for military specifications. The D specification prohibits the use of base stock oils which are considered carcinogenic or potentially carcinogenic. It is safe to assume the MIL-L- 2105D will someday be upgraded to MIL-L-2105E.

Additives

Automotive gear oils contain a sulfur-phosphorus extreme-pressure (EP) additive, which also may include a fatty material for additional lubricity. Although different packages were tried in the past, the sulfur-phosphorus combination is the most common today, since the removal of lead from just about any additive package was mandated.

Industrial gear oil and most EP greases contain this same type of additive package. The big difference is that the automotive gear oils have about twice as much of this EP additive as do the industrial gear oils or greases. The automotive gear oil must have this high level of additive to cope with the severe service in hypoid gearing found in most differentials.

Since this additive package becomes activated with heat, its protection for gears is limited in very low temperature locations. In the arctic, for example, special low pour, highly additized gear oils must be used to protect the gears. The sulfur element activates at higher temperatures than does the phosphorus portion, thereby providing protection over a wide range of temperatures. These elements chemically combine with the surface of the gear teeth to form iron sulfides, which are softer than iron, and will therefore chemically polish the surfaces rather than welding and tearing the surfaces. Oxidation and rust inhibitors are also used to improve performance in service.

Automotive gear oils, designed for service in limited slip or posi-traction differentials, usually contain a fatty additive to ease the slippage required in this type of differential. That fatty additive might be a sulfurized fat to control the corrosive effect of any free sulfur in the gear oil. The use of automotive gear oils and other fluids in manual transmissions is covered in Chap. 7.

Synthetics

The advantages and disadvantages of semisynthetic or synthetic gear oils are the same as stated for semisynthetic and synthetic engine oils. The semisynthetic will suffer from oxidation of the mineral oil portion of the formulation. Fully synthetic gear oils have a big advantage in oxidation resistance and lower internal friction.

Some major differential manufacturers have, and others are considering, extended warranties and extended authorized drain intervals with synthetic automotive gear oils. They also recommend a gear oil by API and SAE standards, e.g., API GL-5, SAE 80W-90.

There can be a notable cost savings with synthetics for a fleet of vehicles which are currently changing out gear oil on a scheduled basis. Those cost savings include less mechanic and driver labor and less truck downtime. Less VI improver, if any, is normally used in a synthetic automotive gear oil because of the natural high viscosity index of the base fluid. Many synthetic automotive gear oils also show SAE grades of 75W because of their excellent low-temperature characteristics so you can see a 75W-90 or maybe even a 75W-140.

Most synthetic products can also show a reduction in oil temperature, easier shifting in manual transmissions, and a definite improvement in cold-weather shifting.

Troubleshooting

Oil-related troubles with automotive gear oil fortunately are not common, but they can occur.

First is oxidation of the base oil. Extended drain intervals, coupled with severe service, oxidize the gear oil beyond its normal service life. A badly oxidized gear oil appears as a black, gummy asphaltic-type deposit, as if someone poured asphalt or tar into the gear case. This condition changes the clearance on the gear teeth and destroys the gears and bearings. Some manufacturers use better-quality base stock oils than do others, so length of service can vary. The price of a gear oil, like so many things, is nothing compared with the cost of rebuilding a transmission or differential.

If there is any doubt in your mind that your gear oil is not giving the performance you need, get a pint sample of the product and send it to the oil laboratory to determine the possible cause of the problem. Poor or nonperformance should not be tolerated and is not cost-effective. Prior to judging performance, confirm that the normal scheduled maintenance has been performed by your personnel.

Second, dirt and/or water can enter the gear case. Dirt, as in an engine oil, makes a very destructive grinding compound. Water can enter a gear case if the application involves driving through water on the job or backing a vehicle into water. Check out any vents in the gear set or transmission to ensure that water and dirt cannot enter. Water contamination will have the "cream in coffee" appearance.

Third, if you are experiencing poor performance from your gear oil, make sure the correct viscosity and the right type of product is being used, and make sure the mechanic is applying the correct viscosity or grade and not mixing up the drums of motor oil and gear oil. Although an engine can live with an accidental change of gear oil, engine oil in a differential or transmission can be disastrous. Get a sample of product with which you have experienced poor performance as well as a sample of the new oil and send it to an oil analysis laboratory.

Try to avoid taking someone's word for something, because mistakes happen all the time. There might have been a temporary replacement mechanic on duty, the mechanic might have had someone else get the oil, or the new product might be contaminated or mislabeled. Murphy's law will prevail!

Self-Quiz

Mark your answers to the questions and compare them with the answers listed at the end of the book.

1. What organization sets the performance standards for automotive gear oils?
   a. SAE
   b. API
   c. ISO

2. What is the current military specification for automotive gear oil?
    a. MIL-L-2104B
    b. MIL-L-2105D
    c. MIL-L-2104E
3. The most common automotive gear oil classification today is:
    a. GL-6
    b. GL-4
    c. GL-5
4. The most common extreme-pressure additive in automotive gear oil today is:
    a. A zinc-phosphorus compound
    b. A sulfur-phosphorus compound
    c. A synthetic fatty compound
5. All mineral oil automotive gear oils meeting the current military specification contain a viscosity index improver additive.
    a. True
    b. False
6. Some manual transmissions do not recommend automotive gear oil.
    a. True
    b. False
7. Use of a synthetic automotive gear oil will void your warranty.
    a. True
    b. False

7

Transmission Fluids

Automatic Transmissions

Passenger Cars and Trucks. The following automatic transmission fluid classifications are obsolete: GM Type A, GM Type A—Suffix A, and GM Dexron. Applications for these fluids are extremely rare, and the Dexron II fluid could probably be used as a substitute in all of them.

Two primary types of automatic transmission fluids (ATF) are used in passenger cars today. Dexron®II and IIE (GM and others) and Mercon (Ford) specifications can be met by a single fluid. Type F (older Fords up to 1977 except for C-6 transmissions) is a fluid similar to Dexron II but with different frictional characteristics. The Ford specification for these C-6 and JATCO automatic transmissions (M2C138-CJ) is now met by many of the Dexron II and IIE fluids. All are dyed red as part of the specifications and to provide assistance in determining the source of an oil leak. Some ATF fluids are available undyed but are primarily used in industrial transmissions in machine tools.

The friction coefficient of Dexron®II/Mercon® fluid *increases* as the sliding speed increases in the transmission. The friction coefficient of the older Ford Type F fluid *decreases* as the sliding speed increases in the transmission.

The Dexron II/Mercon fluid will provide a slower, smoother lockup as the transmission accelerates. The Ford Type F fluid provides a quicker, more positive lockup as the transmission accelerates. If you were to put the GM fluid in a Ford Type F transmission, it would tend to slip during acceleration or when under a load. If you put the Ford Type F fluid in a GM transmission, it would tend to slam into gear.

Both fluids have excellent low-temperature characteristics, high viscosity index, excellent oxidation resistance, and a detergent and anti-wear additive package. Although the fluids have similar properties, they should not be mixed in automatic transmissions.

Some Dexron II/Mercon fluids are qualified against other manufacturers' specifications such as Allison's C-3 and C-4 and Caterpillar's TO-2 and TO-3. Some Dexron II/Mercon fluids were qualified against the Allison C-2 specification but had poor retention of frictional properties in severe service. The newer Allison C-3, and now C-4, specification was developed to include a test for the retention of frictional properties, which was a weak point in C-2 specification.

Although the Dexron II/Mercon products have qualified against the other specifications, the higher cost of ATF, compared with an equally qualified engine oil, will limit its use as a C-3, C-4, or TO-2 and 3 fluid except in cold applications. Transmission fluids may also have an application as power steering fluid in some equipment and hydraulic fluid in cold-weather operations because of its excellent low pour characteristics (see Chap. 8). Synthetic fluids now dominate the power steering fluid market. Specifications keep changing, and Dexron III will probably come out by 1994.

Other Transmissions

Some truck transmissions require an SAE 50 engine oil, SAE 50 gear oil, or straight mineral (non-EP) SAE 90 gear oil. Comparison of these two types of products shows they are very close in viscosity. Although the engine oil contains a detergent-dispersant additive package, it would not be beneficial or detrimental in this type of application. Eaton and Dana recommend engine oil and/or straight mineral gear oil in their transmissions. They do not want a product containing the sulfur-phosphorus extreme-pressure additives because their units contain nonferrous parts which will be stained or corroded by the chemically active additives.

A multiviscosity engine oil such as 15W-40 or 15W-50 should not be recommended for this application. Even the best shear-resistant, multiviscosity engine oil will shear down in viscosity in a transmission and will not provide the protection of an SAE 50 engine oil in warmer temperatures. Caterpillar Tractor is of the strong opinion that heavily loaded gears and bearings will only see the lubricant parameter and film thickness of the base stock rather than the viscosity of the VI improver thickened product. Should cold temperatures make the use of an SAE 50 engine oil impractical, a lighter grade like an SAE 40 or a synthetic transmission oil should be used.

Should you request an SAE 50 "gear oil" from your supplier, it could be confusing as to exactly what product you need. You should specify either an SAE 50 engine oil or a straight mineral or non-extreme-pressure SAE 90 gear oil.

Some major oil companies and private blenders also market synthetic oils which are designed to satisfy both of these requirements. These types of products are beneficial in severe service applications such as trailer towing, extended drain intervals, or cold weather operation. Should you decide to use the synthetic transmission oil, maintenance could probably forgo regular drain intervals because of the superior oxidation properties of the synthetic.

In computing the cost-effectiveness of a synthetic, your figures should include labor, vehicle downtime, driver downtime plus overhead, and the product cost.

Your choice of which product to use, SAE 50 engine oil, SAE 90 non-EP gear oil, or synthetic, should be determined by the other requirements for your vehicles, product stocking requirements, operating conditions of the equipment, and economics. Should you have the choice of the SAE 50 engine oil or the SAE 90 non-EP gear oil, you would probably find the SAE 50 engine oil to be the least expensive and more readily available in many areas of the country. The API service classification, CC, CD, CDII, or CE, is immaterial in this type of application.

One major small truck manufacturer originally recommended an engine oil in its five-speed manual transmissions and experienced excessive gear wear and failures. It changed its recommendation from an SAE 10W-30 engine oil to an SAE 80W-90 extreme-pressure gear oil to solve the problem. Although the manufacturer tried to gain better performance with the use of a lighter oil, it found the loading on the gears exceeded the capability of the engine oil to protect the gear teeth. The lighter oil might have assisted in meeting fuel economy requirements from the federal government. Review of the viscosity charts in the appendix shows an SAE 80W-90 gear oil is very similar in viscosity to an SAE 20W-50 engine oil. The change in the recommendation (1) increased the viscosity over the original engine oil and (2) added an extreme-pressure additive package, thereby increasing the protection of gears in the transmission. It is very important to follow the manufacturer's lubricant recommendation even though it might be updated at a later time!

Some manual transaxles in domestic and foreign vehicles recommend the same engine oil viscosity as is used in the engine. They do not need the extreme-pressure additives found in automotive gear oil and benefit in less fluid friction with the lighter lubricant. Again, it is important to follow the manufacturer's recommendation!

In heavy equipment, Caterpillar and Allison transmissions are very common. They require transmission fluids which meet their most current specifications, such as Allison C-3 or C-4 and Caterpillar TO-2 or TO-3. These specifications are most important in the contracting and mining fields and are easily met by many diesel engine oils as well as other fluids. Qualified engine oils are the most commonly used products to satisfy these requirements because they reduce the error potential of someone adding another fluid to the engine crankcase. In heavy industry, lubricant control is vital to extending the life of heavy equipment.

Multipurpose Tractor Fluids

Every major oil company markets this type of product. It is a unique fluid designed to be truly "multipurpose." It was originally designed to meet or come close to the multipurpose tractor fluid specifications for J.I. Case, John Deere, and International Harvester farm tractors.

These fluids are also usually qualified as Allison C-3 and Caterpillar TO-2 and 3 fluids although they are rarely used for those applications except in colder climates where their low pour points allow for faster fluid circulation.

They serve as a transmission fluid, hydraulic fluid, and gear oil in those applications which call for this type of fluid. The viscosity is in the range of an SAE 10W-20 or 20W-30 with very high viscosity index, outstanding rust and oxidation resistance, low pour point, and good extreme-pressure properties.

Although many multipurpose tractor fluids are recommended by the oil companies as suitable replacements for the Case, IHC, and John Deere fluids, many are not really approved by those manufacturers. The oil companies may say the fluids meet the service fill performance requirements of the manufacturer. That is a fancy way to say they will do the job.

Since equipment manufacturers usually approve only their own fluids and, of course, sell them through their dealers, the oil companies' products are not normally "approved." Each oil company will recommend its products for the applications in which it feels comfortable and will stand behind the performance of its products. Carefully read the product descriptions to determine the applications. These types of products are normally expensive and are normally used primarily in the farming and small isolated mining and construction applications.

Troubleshooting

Fortunately, there are not many problems with transmission fluids. Under normal conditions, transmission fluid will do a good job and have a good service life.

Hauling a trailer will cause automatic transmissions to overheat and require the fluid to be changed out more frequently. If the transmission is completely full and put into severe service without an oil cooler, the fluid can actually boil up and overflow through the fill tube. The addition of a transmission cooler will be a big help but may not completely control the increase in fluid temperature. Too large an oil cooler might not let the automatic transmission fluid (ATF) reach a good operating temperature, so the size of the cooler should be predicated on the service. An oil filter may easily be added to the oil line going to the auxiliary cooler for additional filtration and oil capacity.

There is an ongoing debate regarding the installation of an auxiliary cooler. Some contend the transmission oil should be routed into the main radiator chamber, on into the auxiliary cooler, and then back to the transmission while others contend the main radiator chamber should be bypassed completely, routing the oil line directly from the transmission into the auxiliary cooler and back to the transmission. This routing actually increases the capacity of the cooling radiator to keep the coolant temperature more under control.

If you see automatic transmission fluid which has turned from red to a brownish red in color, it has oxidized and should be changed out immediately. The oxidized fluid may have glazed the clutch plates, which will reduce the efficiency of the transmission until failure occurs.

Working your car back and forth to free it from a snowbank can cause your automatic transmission fluid to be burned from local hot spots in the transmission. Although the red color may darken only slightly, you may notice a burned odor when you check the dipstick.

In manual transmissions, heavy loads and severe service may require more frequent transmission drain intervals. If the fluid is not changed, it will oxidize and varnish the transmission components and shorten the life of the gears and bearings.

In severe or even normal service, regular drain intervals cannot be overemphasized. They will be the cheapest insurance against failure.

Self-Quiz

Mark your answers to the questions and compare them with the answers listed at the end of the book.

1. Dexron is a brand name owned by what automotive manufacturer?
   a. General Motors
   b. Ford Motor Co.
   c. Chrysler
2. The major difference between Dexron and Ford Type F automatic transmission fluid is:
   a. Viscosity
   b. Friction coefficient
   c. Extreme-pressure additives
3. Automatic transmission fluid (ATF) is a very poor low-temperature hydraulic fluid.
   a. True
   b. False
4. ATF cannot qualify as an Allison C-2 or C-3 fluid.
   a. True
   b. False
5. An SAE 50 engine oil is much lighter in viscosity than an SAE 90 gear oil and should not be used in any transmission.
   a. True
   b. False

8
Mobile Hydraulics

Types of Pumps and Oil Viscosities

Gear, vane, and piston pumps are the most common hydraulic pumps regardless of application. Each type of pump has advantages and disadvantages, and one is picked by the equipment manufacturer to perform a particular type of service. Hydraulic pump manufacturers have established a minimum-maximum viscosity range for the proper operation of their pumps (Table 8-1). They have also established a minimum-maximum temperature range for the typical viscosities of common products.

Table 8-2 is a chart showing the minimum-maximum oil reservoir temperatures by type of pump and viscosity grades.

The temperature limitations in Table 8-2 are guidelines for service. Although the upper limits are roughly the temperatures where a typical oil of that type would not have the viscosity to fully protect the pump, the lower limit is only the coolest operating oil reservoir temperature recommended. At a temperature lower than the minimum, the typical oil of that type would be almost too viscous for the pump to move without cavitation and rapid wear. Needless to say, the pump may be started at a lower temperature, but it should not be pressed into service until at least the minimum temperature is attained. If a pump, which has just been started at a lower temperature, is pressed into service, cavitation and accelerated pump wear can occur.

Table 8-1. Viscosity Range, SUS

Type of pump	Minimum	Maximum
Dennison piston:		
Cold starts		7000
Full power	60	750
Dennison vane:		
Cold starts		4000
Full power	60	500
Hydreco gear:		
Cold starts		4000
Operating	70	300
Vickers vane:		
Optimum	80	180

Table 8-2. Recommended Operating Temperature Range by ISO and SAE Grades of Oil

Type of pump	Minimum, °F	Maximum, °F
Dennison piston:		
ISO 32 or SAE 10	55	160
ISO 46	60	180
ISO 68 or SAE 20	75	190
Dennison vane:		
ISO 32 or SAE 10	65	160
ISO 46	70	180
ISO 68 or SAE 20	85	190
Hydreco gear:		
ISO 32 or SAE 10	70	145
ISO 46	80	165
ISO 68 or SAE 20	95	175
Vickers vane:		
ISO 32 or SAE 10	80	143
ISO 46	94	159
ISO 68 or SAE 20	102	177

Lubricants

From these tables, it is easy to see that a lighter viscosity oil should be used when the weather is cold and a heavier viscosity oil when the weather is hot. The temperature ranges are oil reservoir temperatures in operation. The oil reservoir temperature increases over the ambient temperature with operation. Various factors contribute to this temperature rise.

For example, if the ambient temperature is 50°F in the morning and the equipment is cold, the hydraulic oil reservoir temperature might increase by 50 to 60°F while in operation. The oil reservoir temperature is now 90 to 100°F and an SAE 10W oil would be the proper fluid. If, however, the ambient temperature was 100°F and a piece of mining equipment, with four hydraulic pumps running off that reservoir, showed a temperature increase of 80°F, the proper oil would be an SAE 20W-20 or 30. If a lighter or less viscous oil were in use, it would not have the viscosity to protect the pump from accelerated wear. In this case, an expensive pump may only have a service life measured in "days." In some mining and construction operations, the selection of this grade could tie in with the engine oil needed for miscellaneous equipment such as pickup trucks and drills.

In colder climates, consider the use of automatic transmission fluid or a special low-pour antiwear industrial hydraulic oil as an alternate to engine oil as your hydraulic fluid. The red color in the ATF would be helpful in spotting and correcting hydraulic fluid leakage. A dye can be added to other fluids to achieve the same purpose, if desired. Although ATF might be more readily available in some locations, many manufacturers of equipment will recommend a special low-pour, antiwear hydraulic oil over the ATF to protect their equipment.

Although the upper end of the temperature range stated will provide adequate viscosity to protect the pump, the oxidation rate of the oil, at temperatures over 160°F, will increase rapidly. In that case, shortened drain intervals should be a regular part of the maintenance program. The ideal high operating temperature of a hydraulic system is about 120 to 130°F for good lubricant oxidation life.

The key to protecting that hydraulic system is to provide an oil of the proper viscosity at the operating temperature. If the start-up temperature is low, a low-pour or a VI improved oil is proper to cold-start the hydraulic system and still provide pump protection at the operating temperature. A cold hydraulic system should be run at idle for a while to warm the hydraulic oil prior to operation.

In the contracting and mining field, engine oil is the most common product available. Construction jobs and mines want to *minimize* the types of lubricants they stock in order to *minimize* the potential for product misapplication in the equipment. Product consolidation also will allow the operation to purchase lubricants in bulk rather then in drums or smaller containers. The difference in purchase price will be notable over a year's period, and bulk purchasing is much easier to control. The disadvantage to bulk is the cost of the increased inventory and the initial cost of the bulk container.

If an SAE 10W or 20W-20 engine oil is used in the hydraulics, it would not damage the engine should the "hydraulic" oil accidentally be added to the engine as makeup. It would slightly reduce the viscosity of the engine oil but not enough to cause a major problem.

Zinc dithiophosphate and zinc dialkyldithiophosphate is the common antiscuff additive used in both antiwear hydraulic oils and engine oils. Therefore, engine oil will also protect a high-pressure hydraulic pump from rapid wear. The detergent and dispersant in engine oil is of no benefit in a hydraulic application but has no adverse effect on hydraulic performance unless water contamination occurs. Then an emulsion will be formed which will not properly lubricate and protect the pump.

An exception with the contracting and mining companies operating in extreme cold might include a special low-temperature hydraulic fluid especially recommended by the manufacturer for that type of application. This may be a product such as an aviation hydraulic oil, with an extremely low pour point and meeting a military specification, to ensure proper lubrication during start-up.

Although far more attention is paid to the performance of engine oil, a typical contractor or mine will have equal consumption of hydraulic oil and engine oil in heavy equipment. Everyone will track the performance of engine oil by engine life but rarely will anyone track hydraulic pump life. The evaluation of hydraulic performance should be more important in watching expenses in these heavy-duty operations.

Farmers normally use a multipurpose tractor fluid in their hydraulics. It is a product that they will have on hand and is designed to function in tractors which utilize a multifunctional reservoir. Very large farms may utilize large Caterpillar-type equipment, much like a large contractor, so their requirements will have to cover a broader range of products to satisfy the various manufacturers' recommendations.

Some equipment manufacturers utilize pressurized hydraulic systems and excellent filtration to keep dirt out and maintain a clean hydraulic system. If you are going to pull a sample from a pressurized system, vent the pressure prior to removing the fill cap. Better yet, get a maintenance person who is familiar with the equipment to assist you. It is not funny to be sprayed with hot hydraulic oil and has been the source of more than one industrial injury claim.

Care must be taken to ensure that clean oil goes into the system when makeup or an oil change is required. Operations that are dirty should install filtration between the container of oil and the hydraulic reservoir. The area around breathers, dipsticks, and fill caps should be cleaned prior to opening to reduce the chance of dirt contamination.

Nozzles on the oil-dispensing equipment should be wiped clean prior to dispensing product. This protection is even wise for a relatively clean industrial plant.

A common oversight, particularly in mines or on construction jobs, is to remove the small drum bung vent so that the drum will not build up a vacuum when the oil is removed. This opening will allow dirt to enter the drum and contaminate the new oil. Special vented bungs or air breathers are available and should be used to replace the small bung. This will prevent dirt contamination and allow the drum to breath as new product is removed from the drum.

To not allow the air to replace the oil withdrawn invites an industrial injury. One example of this occurred several years ago in an automotive shop. A simple hand pump was being used to remove oil from a 55-gallon drum. The mechanic felt there was an increasing resistance to pumping the oil when suddenly the drum imploded, jumped about 6 inches into the air, and landed back on the mechanic's toes. This kind of industrial accident can and should be avoided by properly venting the container.

Self-Quiz

Mark your answers to the questions and compare them with the answers listed at the end of the book.

1. Hydraulic pump manufacturers set minimum viscosity limits for their pumps but do not set maximum viscosity limits for cold starts.
   a. True
   b. False
2. The maximum operating temperature of any pump should not exceed 170°F because of the seal material.
   a. True
   b. False
3. The most commonly used hydraulic fluid in the fields of contracting and mining is:
   a. Antiwear hydraulic oil
   b. Engine oil
   c. Turbine-quality hydraulic oil
4. The majority of farm tractors use the following product in their hydraulic system:
   a. Engine oil
   b. Multipurpose tractor fluid
   c. Antiwear hydraulic oil

5. The following type of additive does an excellent job of protecting high pressure hydraulic pumps:
   a. A zinc-phosphate compound
   b. A sulfur-phosphorus compound
   c. A synthetic fatty compound
6. An SAE 30 SG engine oil is an approved replacement for multipurpose tractor fluid in warmer climates.
   a. True
   b. False
7. The concern with running a hydraulic pump hotter than the recommended temperature for a given viscosity grade is:
   a. The zinc additive will become corrosive.
   b. The hydraulic motion will increase.
   c. The oil will be too thin.

9

Greases— Automotive and Industrial

Principles

The first grease was probably animal fat used to lubricate the axles on chariots or wagons. Grease is the most magical and mystical part of lubrication because grease actually does not lubricate. The function of lubrication is accomplished by the oil or other lubricating fluid which bleeds from the grease. If the grease did not bleed the lubricating fluid, there would be little or no lubrication. Grease is little more than "thickened" oil or other lubricating fluid.

Sometimes a maintenance person opens a pail or drum of grease and notes a puddle of oil on top of the grease. Many think the grease is old and is "breaking down" and voice a complaint about it. This is not the case at all. The grease is simply releasing the oil as it should, and that small puddle of oil is completely normal.

Grease is a combination of lubricating fluid, a thickener, additives, and in some cases, fillers. The thickener selected will determine the basic characteristics of the grease. The lubricating fluid is selected to perform a particular function or be multipurpose and the fillers and additives are selected to enhance the performance in difficult applications.

Grease is basically used in applications in which an oil lubricant would leak out, could not properly seal the lubricated part, and would fall away or not reach the point of application. Grease is used where oil will not stay.

Greases are classified automotive and/or industrial by manufacturers based on (1) viscosity of the stock oil, (2) additives, and (3) thickener used. Some oil companies manufacture the same grease, under two different names, classifying one as automotive and one as industrial. This is done only for accounting and advertising purposes. A semi truck driver would tend to use a grease that is labeled "heavy-duty truck bearing grease" while that same grease might be fully suitable for a multipurpose industrial grease and might even be available as "high-temperature multipurpose industrial grease." Your choice of grease should be based on the properties of the thickener and the viscosity of the oil in the grease. The viscosity of the oil should be chosen the same way as a lubricating oil is chosen as stated in Chap. 2.

Greases are made by various methods; some are cold-made, some in open or closed kettles, and some in continuous grease makers. Products made in a continuous grease maker will have a cost advantage over products made by other methods.

Pricing on grease is based on (1) how new the thickener is on the market, (2) competition, (3) advertising of the product, (4) type and viscosity of the oil used, (5) additives involved, (6) packaging, (7) the method by which the grease is manufactured, and (8) competition. The smaller the container, the higher the price. You will not find, however, that the savings of buying grease in bulk loads and then applying it with a grease gun is cost-effective because of the handling time and expense. Bulk grease is very cost-effective when the bulk grease can be automatically fed into large-volume applications, as in a steel mill's rolling mill or other large-volume grease users.

The use of grease is advantageous in the following types of applications and for the following reasons:

1. It decreases the dripping and spattering of the lubricant used by acting as an additional seal to reduce leakage.

2. It decreases the frequency of lubrication is some cases. Grease will stay in an application where some oils may leak out.

3. It helps seal out contaminants such as water and dirt. In corrosive atmospheres, it helps seal out those corrosive elements.

4. It hangs on the application in an intermittent operation where oil would drain away.

5. It can suspend solids, such as moly, graphite, and zinc oxide, much

better than oil. In certain applications, the addition of solids can be beneficial.

Metallic Soap Thickeners

Historically, many types of thickeners and methods have been used to make grease. The following are the most common types used in the industrial and automotive market today. They are meant to show examples of how some greases are made and are not the only or the latest way grease is manufactured.

Aluminum Soap Grease

An aluminum distearate soap is mixed with the proper proportions of oil. It is blended at room temperature and then heated to about 280 to 300°F until the oil and soap are completely blended. The remainder of the oil is then added. It can then be poured into pans for cooling, where it forms a firm cake. Fast cooling results in a firmer product and slow cooling results in a softer product. It is then worked and milled for uniform consistency. Aluminum greases are very tacky but never fibrous. They are not used in industry today.

Hydrated Calcium Soap

The basic fatty ingredient is tallow which may come from horse, hog, beef, or mutton fat to make fatty acid. An alkali, calcium hydroxide, is added to react with the fat to make the calcium soap. Water is added and the mixture is heated. Oil is added as the mix cools, along with some water, to make the finished product. This grease has been called "lime" soap grease because calcium hydroxide or, by another name, "hydrated lime" was used in the process.

Hydrated calcium greases have a typical dropping point of around 200°F. Above approximately 170°F, the water used to make the grease will start to be driven out of the mixture and the grease structure will be destroyed. The grease has very good water resistance and fair to good work stability in service. Low-temperature handling is very good unless a very viscous oil is used in the formulation.

Manufacturing costs are low for this calcium grease since it utilizes inexpensive components. This grease is rare in industry today because of its limitations compared with more modern greases.

Anhydrous Calcium Grease

The most successful formulation of this water-free calcium soap grease involves the use of 12-hydroxystearic acid. Reacting this acid with lime in the presence of oil eliminates the need for water.

The heat resistance is somewhat better than that of hydrate calcium soap grease with a dropping point of about 280 to 290°F, but it is still lower than greases made with other types of thickeners. Low-temperature characteristics are satisfactory and shear stability is good. They are water-resistant, but rust protection must be enhanced by additives.

Manufacturing costs are relatively high when compared with lithium-based greases, which offer much better high-temperature properties.

Anhydrous calcium grease enjoys limited use in industry. Its present appeal is based mostly on its natural tackiness, water resistance, and better high-temperature resistance. Some small companies have turned it into a specialty grease by using a very viscous lube oil, fatty compounds, extreme-pressure agents, and other additives.

Sodium Soap Grease

Reacting sodium hydroxide with fats and fatty acids makes sodium or soda soaps. It is slowly heated with some water similar to the hydrated calcium process. As the soap and oil are mixed, it segregates into tough ropelike fibers. This is called a fibrous or stringy grease.

Sodium soap grease structure varies with the manufacturer. This variance is caused by the type of fat and fatty acid used as well as the proportions. The mixing of the components, modifiers, milling, and cooling will also cause a variance in structure.

Sodium greases have a dropping point over 300°F and a top service limit of about 250°F, are fibrous, and have only poor water resistance. In the presence of small amounts of water, they draw the water away from the metal, form an emulsion, and protect metal surfaces from rust. Low-temperature characteristics are hampered by the fibrous texture of the grease. They are in very limited use in industry today.

Lithium Soap Greases

Lithium soap greases came into their own during World War II as the first real multipurpose grease. Fat, fatty acid, some oil, and lithium hydroxide are heated together to form a lithium soap grease base. The most advanced method for making lithium grease is with 12-hydroxystearic acid. This method is the most dominant method used today.

Lithium soap greases have very good high-temperature properties with a dropping point around 365 to 375°F and an upper service limit of about 275°F. They have good water resistance and excellent mechanical stability. Low-temperature handling is very good. Lithium grease can absorb up to 50 percent of its weight in water and still lubricate and protect the metal surfaces from rust. It is a very good multipurpose grease and is commonly found in industry today. Lithium soap greases are the leader in industrial and automotive grease sales.

Without expressing an opinion on quality, these major oil products, and others, trademarked by their manufacturer, generally fall into this class of product:

Chevron Dura-Lith

Mobilux

Shell Alvania

Texaco Marfax and Regal AFB

Unocal MP Automotive and Unoba F

Complexed Metallic Soap Thickeners

The complexing of thickeners has given most grease families a relatively new high-temperature capacity. All the complexed greases are high-temperature greases because that is the basic nature of the "complexed" soap which is used as a thickener.

There are two ASTM test methods of measuring the drop point of a grease: D566, which has an upper limit of 500°F, and D2265, which has an upper limit of 625°F. These very high drop points are great for advertising but the high-temperature greases with mineral oil are still restricted to an upper service limit of about 300°F.

At elevated temperature, the oil's oxidation rate will be extremely high and regreasing intervals will be extremely short. Another point in very high temperature applications is the viscosity of the base oil. Again, oil does the function of lubrication. At high temperatures, that base oil will be extremely thin and may not properly lubricate the parts to be protected. For that reason, the base oil in a high-temperature grease, which is used in high-temperature applications, should be far more viscous than that used in a conventional industrial or automotive grease.

Aluminum Complex

No major improvement is found here except for a higher drop point. It has been observed that this complex grease, after being subject to high temperatures, can turn rubbery upon cooling.

Low-temperature characteristics are fair to good. Shear stability is good to excellent. While water washout and some other water tests show only good results, water spray resistance is considered excellent. Aluminum complex greases, in general, are not in common use in industry or the automotive field.

Without expressing an opinion on quality, these major oil products, and others, trademarked by their manufacturer, generally fall into this class of product:

Chevron ALTMP

Sunaplex

Unocal Aluminum Complex Grease 9998

Unocal Low Temp

Calcium Complex

This type of grease was pushed by some major oil companies seeking a high-temperature multipurpose grease not using a clay-type thickener. Calcium complex greases lacked the ability to be used in long grease lines. It would solidify in the lines and could not be removed. The lines had to be replaced. It also hardened if kept under pressure for a long period of time.

Low-temperature characteristics and water resistance are satisfactory. The grease requires a large amount of thickener for proper consistency. Although they enjoy a small degree of acceptance, calcium complex greases have basically been obsoleted by other types of high-temperature greases.

Without expressing an opinion on quality, these major oil products, and others, trademarked by their manufacturer, generally fall into this class of product:

Chevron Heavy Duty

Mobilplex

Barium Complex

This thickener is not the normal barium soap but rather a barium complex soap using acetic acid as a complexing agent, as do the calcium complex greases. This grease requires a large amount of power to manufacture since it becomes very firm and stringy during manufacturing.

These characteristics have made it a choice factory fill grease at some American car manufacturers. It has not gained much acceptance in the industrial market or even the automotive aftermarket.

Lithium Complex

This grease has taken all the good properties of lithium grease and added higher-temperature performance. Its multipurpose capabilities make it a very good all-around grease thickener for both industrial and automotive greases. By adding different types and viscosities of lubricating fluids and additives, these multipurpose capabilities are further expanded.

Lithium complex greases could easily become the next leader in multipurpose grease lubrication. Without expressing an opinion on quality, these major oil products, and others, trademarked by their manufacturer, generally fall into this class of product:

Mobilgrease HP

Mobilith

Texaco Starplex

Nonsoap Thickeners

Organo-Clay Greases

The clay chosen must be closely matched with the oil to be used. Grease production requires thorough dispersion of the clay in the oil. One method is to add about one-third to one-half of the oil to the clay and mix until uniform in texture. A special pregelling agent, often called a polar dispersant, is added as the mixing continues. Additives and the remainder of the oil are added and then the grease is milled to the desired consistency.

The low-temperature properties of clay greases are satisfactory. When the grease contains a high-viscosity oil for improved high-temperature performance, the low-temperature properties tend to be much poorer.

High-temperature properties of this type of grease are very good, with a dropping point of over 500°F. Since the thickener does not melt, clay greases can be used in applications in which temperatures could hit 500°F. Relubrication is required after a few hours because of the short service life of the oil in the hot application.

Without expressing an opinion on quality, these major oil products and others, trademarked by their manufacturer, generally fall into this class of product:

Mobiltemp

Shell Darina

Polyurea Grease

Polyurea greases are manufactured by mixing oil with suitable amines with an isocyanate or a diisocyanate being added after mixing. Some heat is generated in the reaction but not the by-products, such as water or methanol, which can occur in some other grease processes.

These greases are comparable with some of the other high-temperature complex greases. They are very effective in the lubrication of hot ball bearings. Low-temperature handling is satisfactory. Water resistance is satisfactory and, in some cases, excellent. Compatibility with other greases can be a problem.

Some grease experts have stated that the work stability of a polyurea grease is relatively poor. They do perform very well in low-shear applications such as ball bearings, but not as well in high-shear applications such as roller bearings. Some experts feel a different formulation would perform better in the high-shear-type applications.

Without expressing an opinion on quality, these major oil products and others, trademarked by their manufacturer, generally fall into this class of product:

Chevron Polyurea

Chevron SRI

Lubricating Fluids
Mineral Oils

Mineral oils are the most common lubricating fluid used in grease. Different types of mineral oil are used to satisfy the needs of a variety of functions. The lubricating fluid should be matched up with the properties of the thickener used for satisfactory performance. If the grease is for a high-temperature application, a high-temperature thickener would be used, preferably with an oil of higher than normal viscosity.

A high-viscosity, good-quality oil would be proper for higher-temperatures applications which would require extended service. The oil would thin, with heat, to the proper viscosity needed to properly lubricate and would have the rust and oxidation resistance to perform over a longer time.

Low-quality viscous black oils can be effectively used in heavy-duty, short-duration-type applications such as a steel mill or fifth-wheel grease. In these types of applications, grease is continuously applied to a bearing surface, so oxidation life of the oil is not a major factor in performance. Use of this type of oil can reduce the cost of the grease so that it's competitive in that type of market. This type of grease would be perfect for short-term, rough-type applications.

For a grease designed for low pressures and high speed, such as a high-speed spindle bearing or motor bearing, a light-viscosity, rust- and oxidation-inhibited oil would perform very well. The rolling elements of that bearing cannot plow through a viscous oil without causing excessive heat. To use a grease containing a very viscous oil, like an ISO 460 oil, could cause a failure because the proper oil film could not be established.

Multipurpose greases typically use an ISO 100 or 150 grade lubricating oil for all-around good performance. Some manufacturers even go as high as an ISO 220. These oil viscosities would be typical in a multipurpose industrial or automotive grease.

Many grease manufacturers make a semifluid grease. It is part of a particular thickener line and may contain either a more viscous oil or a less viscous oil than the other grades in the line. The semifluid greases, like an NLGI 00 or 000 grade, perform well in leaky gearboxes that are not high speed or do not have a circulating oil system. They cling to the gear teeth like honey. They can also perform well in large plain bearings with a timed grease injection system to replace a viscous oil which would tend to leak.

If that multipurpose grease were made in an NLGI grade 0 or 1 and designed for lower temperatures or a central system with long grease lines, a lighter-viscosity oil would commonly be used. The same is true for a grease labeled "ABC Grease Arctic." This grease is designed to flow or be pumped at lower temperatures. A viscous oil would reduce this ability. Some greases use a naphthenic oil for low-temperature applications because of its good natural low pour point. That same grease, however, might not function well at elevated temperatures.

Remember, the oil in the grease does the lubrication, and that oil should be selected to form the oil wedge discussed in Chap. 2.

Every customer would like to use only one grease in the plant. Although this is possible in some plants, it is not practical in other plants where grease applications are vastly different. Each application in the plant should be weighted and the grease recommendations made accordingly. If the company wants to compromise, that can be its choice. If the plant chooses to consolidate, it should be advised by its supplier that this is compromise, and the more severe applications should be closely monitored.

The final selection and quality of mineral oil to be used in a grease are predicated on the following:

1. Length of expected service—20 minutes or 20 months
2. Expected temperature range of the application—cold or hot
3. Speed of the application—slow or fast
4. Cost-effectiveness of the product for the applications

Other Fluids

A variety of other fluids, loosely called synthetics, are made into grease for either general or specialized applications. They include fluorinated hydrocarbons and ethers, which have a high resistance to oxidation, and polyphenyl ethers, which are radiation-resistant. For general multipurpose use in industry, they are not considered cost-effective. However, for specialized applications under severe conditions they have the potential to produce savings or cost avoidance far in excess of the increased product costs.

Some of the more common "synthetic" products follow.

Polyglycols.   These are formed by polymerizing ethylene glycol and/or propylene glycol into fluids. Under high-temperature conditions, they decompose into volatile compounds, leaving no residue. There may, however, be a residue from the thickener used. They are very good for high temperatures but poor for wet applications. They also tend to soften paint, varnish, and other finishes.

Organic Esters.   These are formed from dibasic acids mixed with monohydroxy alcohols and other methods. They are used for low- or wide-temperature application such as aircraft lubricants but they will soften paint, etc.

Silicones.   These do not contain carbon atoms but rather a chain of alternating silicon and oxygen atoms. They are useful in high-temperature and wide-temperature-range applications but do not respond well to most load-carrying additives which help mineral oils and greases perform. They are very costly and are normally used just in specialized applications.

Synthetic Hydrocarbons.   These are becoming more common in special and multipurpose applications in industry today. Two basic types are used; (1) polyalphaolefins (PAO) and (2) alkylated aromatics.

PAO stocks are usable over a wider temperature range than mineral oils; they are less volatile and more heat-resistant and oxidation-stable at high temperatures. They can be used with different thickeners. They can be expensive, but they make superior greases for severe applications.

Alkylated aromatics are a much different chemistry from the PAO stocks. Although their properties are similar, the viscosity index of the alkylated aromatics is lower, resulting in more viscosity change with temperature.

Additives, Fillers, and Dyes

Most extreme-pressure (EP) greases contain the same type of sulfur-phosphorus compound that is used as an EP agent in industrial or automotive gear oils. The amount added, however, varies with each type of product.

Some grease manufacturers add one of the zinc antiscuff compounds which is used in engine oil and AW hydraulic oil. The purpose, with both types of additives, is to gain some protection against shock loading or extreme pressures on a bearing surface.

The use of an EP grease is acceptable, even in bearings not subject to shock loading or extreme pressures, and is typical of a multipurpose grease. The presence of the EP agent is not detrimental in this type of application. The exception would be electric motor bearings, where a grease containing EP agents is not preferred.

Other additives might include moly (molybdenum disulfide), graphite, and other soft metals or minerals. Moly is probably one of the best known and most used today. Concentrations of 1 to 3 percent or more in grease are common on the market today. Moly and graphite reduce friction because of their low internal friction and by filling in the small voids in the bearing surfaces. Moly, because it is a sulfite compound, can become corrosive in a bearing at high temperatures, just like sulfur in an EP gear oil or corrosive cutting oil. This temperature limitation is of no concern since that temperature is well above the dropping point of most greases. Both moly and graphite provide excellent protection in applications which have sliding or oscillating motion. Gray is the normal color of a grease which has had moly or graphite added.

Use of a moly or graphite grease in a seldom-serviced antifriction bearing can leave a cake of concentrated thickener and moly behind in the bearing after the oil has bled out. Unless that bearing is completely flushed out when serviced, a bearing failure can occur from a buildup of the thickener and additive residue.

The additive, Paratac, is added to some greases to give them a "tacky" feel and assists the grease in sticking to the application. Paratac will, however, shear or break down during use, and its benefits are relatively short-lived. Many old-time mechanics were used to the tackiness of the now obsolete long fiber wheel bearing grease, so the addition of this additive gives them confidence in the new grease's ability to perform.

Powdered zinc and copper are sometimes used in greases designed for drill rod thread lubricant. The metal is actually suspended in the grease. As extension rods are added to extend a drill bit, the threads must be lubricated with an "antiseize" agent so they will not weld together under the extreme pressures. The zinc serves that function very well. Since the powdered metal adds a great deal of weight to the product, the normal volume of 1 pound of a conventional grease would probably weigh 5 pounds in a zinc-type product.

Dyes and other fillers can be added, but many are primarily for cosmetic and marketing purposes. Some customers just fall in love with a red or white grease while others prefer purple. The addition of moly or graphite will turn the grease gray in color. In some industries, a white grease is preferred so any grease that may drip on the finished product will be less noticeable.

Selection of a Special Grease

Looking at some special cases, we will try to fit the grease to the application.

1. High-temperature applications can be solved by the choice of a high-temperature thickener. It is even better if we can also combine this thickener with a more viscous oil so it will have adequate viscosity at the elevated temperature to protect the bearing surfaces.

2. Very wet applications can be handled by using a grease thickened with an anhydrous calcium thickener. This thickener is as close to

waterproof as we can get. If there is only casual water contact, a lithium soap grease, being water-resistant and having the ability to absorb some water, would do quite well. The use of a "waterproof" grease can leave the bearing surfaces exposed to free water. Many water-resistant greases combine with up to 50 percent of their weight in water and may assist in reducing corrosion.

3. Very low temperature applications can be handled by using a smooth-texture grease like lithium combined with a low-viscosity, low-pour-point mineral oil or synthetic fluid. A softer grease like NLGI 0 or 1 would assist the movement of the grease in or to the application.

4. Very high temperature applications can be handled by a grease thickened with nonsoap or complex thickener and containing one of the synthetic fluids. Regular mineral oil would oxidize very rapidly at these high temperatures. A synthetic fluid would resist oxidation better than a mineral oil.

5. Heavily loaded applications can best be handled with a viscous mineral oil or PAO combined with moly and EP additives. If the temperatures are normal, a lithium thickener will do the job very nicely. If the temperature is high, a lithium complex or nonsoap grease would do better.

6. Very high or low speed applications can be handled by a grease with a low-viscosity oil for high speeds and high-viscosity oil for low speeds. Remember: high speeds—light oil; low speeds—heavy oil.

7. In general, most normal grease applications would probably best be satisfied by a lithium complex, EP grease with an ISO 150 or 220 mineral oil in it. It would be cost-effective and versatile for over 95 percent of the applications in industry, contracting, and mining.

Self-Quiz

Mark your answers to the questions and compare them with the answers listed at the end of the book.

1. A small puddle of oil on top of the grease in a newly opened pail of grease indicates the grease has separated and is not suitable for service.
   a. True
   b. False

2. The type of thickener used to manufacture a grease will determine if it is a high-temperature grease.
    a. True
    b. False

3. What is roughly the dropping point of a lithium soap grease?
    a. 255°F
    b. 325°F
    c. 370°F

4. Lithium soap greases entered the marketplace in the late 1960s and were quickly replaced by the line of hydrated calcium greases.
    a. True
    b. False

5. The viscosity of the oil in the grease has no bearing on applications because the grease does the lubrication.
    a. True
    b. False

6. Purple grease is an indication that extreme-pressure additives have been added to the grease.
    a. True
    b. False

7. Gray grease is an indication that moly or graphite has been added to the grease.
    a. True
    b. False

10

Industrial Hydraulics

Introduction

The principles in industrial hydraulics are the same as in mobile equipment hydraulics—the transmission of power, through a liquid, to the workplace. Water can act as a hydraulic fluid as can a water-oil emulsion, glycol-water mixture, oil, or almost any other fluid. Each type of fluid has its advantages and its disadvantages.

Water-oil and glycol-water are limited in operation by temperature; other fire-resistant fluids are too expensive unless necessary, and just plain water will cause rusting. Petroleum oil is the most commonly used medium because it has more advantages than disadvantages. Oil is the most cost-effective medium in normal industrial hydraulic applications.

Types of Pumps

The same types of hydraulic pumps are used in industrial equipment as in mobile equipment applications. The same pumps have the same fluid and temperature requirements. The biggest difference in industrial applications is temperature range. In an industrial plant, the ambient temperature is normally constant or fluctuates within a narrow temperature range.

Many machine tools and other industrial hydraulic systems have heat exchangers or coolers to control the temperature of the hydraulic fluid in an ideal range for proper operation and long oil life. Proper operating temperature of the hydraulic fluid is critical in many numerically controlled machine tools. A variance of only 20 percent in viscosity can adversely affect the machine's operation.

Hydraulic oil reservoir temperatures should be near 120 to 130°F for long oil life. If you keep the oil clean and dry and the temperature within that range, you can expect the service life to be measured in years rather than months.

Since the industrial and mobile equipment pumps have the same requirements, the viscosity and temperature charts listed in Chap. 8 are repeated here as Tables 10-1 and 10-2 for your information.

Plants which have a wide variety of hydraulic fluid requirements will find it beneficial to consolidate most of their requirements into an ISO 46 AW grade. This will allow them to minimize their stock of the ISO 32 and ISO 68 grades, keeping them for special critical applications requiring those grades.

This consolidation will also allow the main hydraulic oil to be received in bulk rather than in smaller containers with the appropriate cost savings. Cost savings include drum deposits, cost of receiving and handling of the drums, returning the drums, and changing pumps in each container used.

Table 10-2 illustrates the operating temperature range of hydraulic oil by ISO grade, SAE grade, and pump type.

Table 10-1. Viscosity Range, SUS

Type of pump	Minimum	Maximum
Dennison piston:		
Cold starts		7000
Full power	60	750
Dennison vane:		
Cold starts		4000
Full power	60	500
Hydreco gear:		
Cold starts		4000
Operating	70	300
Vickers vane:		
Optimum	80	180

Table 10-2. Recommended Operating
Temperature Range by ISO and SAE Grades of Oil

Type of pump	Minimum, °F	Maximum, °F
Dennison piston:		
ISO 32 or SAE 10	55	160
ISO 46	60	180
ISO 68 or SAE 20	75	190
Dennison vane:		
ISO 32 or SAE 10	65	160
ISO 46	70	180
ISO 68 or SAE 20	85	190
Hydreco gear:		
ISO 32 or SAE 10	70	145
ISO 46	80	165
ISO 68 or SAE 20	95	175
Vickers vane:		
ISO 32 or SAE 10	80	143
ISO 46	94	159
ISO 68 or SAE 20	102	177

Engine Oils

The oxidation life of a motor oil is less than that of a quality rust- and oxidation-inhibited hydraulic oil. Because of the high level of additives, it usually costs as much or more than a good-quality antiwear hydraulic oil.

Engine oils are completely suitable for industrial hydraulic systems. They are not, however, the best products to use. The detergent additive package will cause any water to become emulsified in the oil and circulated rather than settling out. Any water-based coolant will also completely emulsify in the oil. Oil, free of emulsified water, will protect high-pressure hydraulic pumps much better than a water-in-oil mix.

Engine oils are more properly used in contractor and mining hydraulics where other engine oils are used, and should an engine oil used for hydraulics be added to an engine, no major harm would be done.

Common Types of Industrial Hydraulic Oils

The two most common types of industrial hydraulic oils are (1) rust- and oxidation-inhibited mineral oils and (2) rust- and oxidation-inhibited mineral oils with antiwear additives.

Rust- and oxidation-inhibited mineral oils (R&O) were formerly the most widely used hydraulic fluids. Many were turbine-quality mineral oils and had excellent oxidation life even under adverse conditions. The additives increase the oxidation life over that of the excellent base stocks used in the typical products.

Many products advertised TOST (turbine oxidation stability test) results over 3000 hours. They were, and many still are, excellent products. High TOST results do not necessarily relate to excellent service in a hydraulic system. They are test results under given conditions.

As the families of hydraulic pumps increased in pressure, pump wear rates increased with R&O oils to a level unacceptable in industrial applications. Contrasting to this was the acceptable pump life in mobile equipment hydraulics, with the same type of pumps, which commonly used engine oils. Experiments showed that the antiscuff compounds commonly found in the engine oils were the reason. The antiscuff additives protected the pump components better under the higher pressures.

Zinc antiscuff or antiwear additives were added to R&O hydraulic mineral oils as a result of the above experiments. Today, antiwear, or AW, hydraulic oils are the dominant hydraulic fluid in industrial applications. They are designed to protect the pump components operating at high pressures of over 1500 psi. Although the TOST results are typically lower than for straight R&O oils, many products have shown good to excellent longevity in service.

The biggest difference in AW hydraulic oils is the quality of the base stock used. Some oil companies manufacture products using the cheaper base stocks and try to beef up the oxidation rating with additives. This increases the test numbers, but when the oxidation inhibitor is consumed, the oxidation rate of the base oil skyrockets. When this increase occurs, the hydraulic oil will plug filters and deposit sludge and varnish in the hydraulic system. Others use quality base stocks plus additives so the oxidation rate increase is slower after additive depletion and the fluid can be changed out before sludge or varnish forms in the hydraulic system.

Since the AW hydraulic oils are the most common now, steps had to be taken to expand the service capabilities of this type of fluid.

For outside service or for cold service in a freezer or commercial cooler, pour-point depressants can be added to reduce the pour point of the oil. A typical paraffinic stock oil AW hydraulic oil would have a pour point of about 20°F without a pour-point depressant and about −10°F or less with one. Although this does not change the viscosity requirements of the hydraulic pump in the system, it does allow the fluid to move upon cold start-ups and then warm in mild service until reaching operating temperature.

A viscosity index improver can also be added to expand the service temperature range of an oil that is used in a broad ambient temperature range. Since a lighter base stock oil would be used in the finished product, a further reduction in pour point would be normal. In essence, a "multiviscosity" hydraulic oil would be created with the same advantages and disadvantages as those of a multiviscosity engine oil.

The life of the VI improver should be monitored to ensure it has not been subject to damage by the shearing action of the hydraulic pump and components.

The performance of any hydraulic fluid depends largely on the good routine maintenance of the hydraulic systems. A system that is allowed to run very hot will oxidize any oil, some faster than others. Allowing water and coolant to remain in a system will cause shortened component life. The hydraulic fluid must be kept clean, dry, and in a reasonable temperature range for long life.

PROPER MAINTENANCE EQUALS PERFORMANCE!

Other Fluids

Automatic transmission fluids (ATF) are sometimes specified by a machine manufacturer as the hydraulic fluid. This is done because ATF has good protection for the hydraulic components, is dyed red so leaks are easy to identify, and assures the machine manufacturer that a good hydraulic fluid is being used.

They want this assurance because some lower-quality AW hydraulic fluid is on the market today. Machine manufacturers do not sell a "machine"; they sell "results." To achieve those results, that machine needs a quality hydraulic fluid and ATF will provide it, if specified by the manufacturer.

Internal clutches in the machine drive will be another reason ATF may be recommended by the manufacturer. If the manufacturer recommends ATF, do not substitute another product.

Some very large leaky extrusion presses will use water with 1 to 3 percent water-soluble coolants in the hydraulic system. The soluble oil will provide some lubricity and rust protection. When the leakage is excessive and the leaks are almost impossible to control, this method has proved to be quite acceptable and cost-effective. It is an excellent combination of cost-effectiveness and fire control in that order.

The waste emulsion can be broken into oil and water by different means, even adding acid to the emulsion. The oil is removed and disposed of as normal oily waste and the water goes down the sewer after being properly treated. Some localities monitor this method very closely for pollution violations.

Fire-Resistant Fluids

A very important thing to remember is that fire-resistant hydraulic fluids are just that; they are *fire-resistant*, not *fireproof*. They will ultimately burn. Fire-resistant fluids will give the machine operator a chance to escape out of the area and/or start a fire suppression system. Three basic types of fire-resistant (FR) fluid are used in industry today. They are (1) inverted water-in-oil emulsions, (2) glycol and water, and (3) phosphate esters.

Water-in-oil inverted emulsions are parts of water surrounded by parts of oil. The outer oil surface protects the machines from excessive wear and rust. Spray this product on a source of ignition, like a red-hot steel bar, and the water will break out of the mix and turn to steam. This will blanket the point of ignition with steam and snuff out the fire. When the water phase is gone, the oil residue will burn if the source of ignition still exists.

The water content in water-in-oil products is very critical. Considering that about 45 percent water is normal, as the water phase evaporates off the fluid and the percentage decreases below about 40 percent, the emulsion will decrease in viscosity and have greatly reduced fire suppression. Adding straight water over this 45 percent will increase the viscosity of the mixture. The water content of the mixture should therefore be closely monitored and maintained in the 42 to 48 percent range.

Water-in-oil emulsions are the least expensive of the FR fluids and are compatible with common seal material such as Viton and Buna-N (nitrile) rubber. Do not use butyl rubber, leather, natural rubber, or cork composition materials. Fabric-reinforced packing will absorb water and will swell when in contact with this fluid.

Because of the water phase in the inverted emulsion type of product, the top service temperature of the fluid will be approximately 130°F. Again, the water content of the fluid is critical and must be monitored and closely maintained for satisfactory performance.

With glycol-water fluids, controlling the water content is not a major problem as it is with water-in-oil fire-resistant fluids. Price, reliability, and ease of maintenance make glycol-water fluids a good general choice for a fire-resistant fluid.

Water-glycol fluids are more stable and tend to provide more lubricity than the water-in-oil products. Properly inhibited water-glycol products are compatible with all metals except zinc and cadmium. They are more expensive than the water-in-oil products and perform their fire suppression in the same way. The water phase turns to steam and snuffs the flame. When the water phase is gone, the glycol residue will burn. Seal material recommendations are the same as for water-in-oil fluids.

Phosphate ester type fluids are chemical in nature and provide more stability and lubrication equal to petroleum hydraulic fluids. Although they are the most expensive of the FR fluids, they can also be the most maintenance-free fluid in many applications. Special seals are required when a phosphate ester fluid is used. If a machine is designed to use FR fluids, it will normally come from the factory with special seals. Butyl, silicone, Viton, PTFE (Teflon), and certain other seal materials are generally usable. Phosphate ester fluids are not compatible with most paints or pipe compounds.

Phosphate ester fluids are very fire-resistant but will burn when the fluid-air mixture is just right. They tend to be the most fire-resistant of all the fluids, but test results will vary because each test is run differently. They are still only fire-resistant and not fireproof.

Major oil companies will normally purchase the water-glycol and phosphate ester fluids and rebrand them under their company names. Follow all the limitations and instructions from the manufacturer or marketer of the fluid. Proper installation of the fluid, seals, painted surfaces, and cleanliness of the hydraulic system are critical in all cases.

Other synthetic products and synthetic blends are used in some hydraulic systems where the additional cost can be justified. Polyglycol-synthetic blends can give superior performance in hydraulic systems, particularly where petroleum distillates such as natural gas, propylene, and other hydrocarbon might be involved. Polyalphaolefin (PAO) stocks have proved to be superior hydraulic fluids capable of service in severe applications where high and/or low temperatures are involved. Their naturally high viscosity index will provide a broader operating range than that supplied by mineral oil AW hydraulic oils. Their natural properties will also provide greater wear protection for the pumps and cylinders, much longer service life, and excellent opportunities for reclaiming and reuse. It is, however, mandatory that hydraulic system leakage be under total control to achieve the proper cost-effectiveness. A user cannot afford external leaks of the hydraulic fluid. A small leak of 1 drop every 10 seconds can cost a user in the area of $800 per year in lost fluid.

Filtration

In addition to proper hydraulic fluid temperature control, there should be efficient filtration which can be achieved through in-line and/or portable filters. Most informed people feel that 10 micron filtration is adequate and 5 micron is even better depending on the application. In systems which supply small clearance control valves, finer filtration, as low as 0.5 micron, is mandatory. The size of this type of filtration is usually mandated by the equipment manufacturer. Numerically controlled machine tools and turbines will normally require the finest degree of filtration.

The purpose of filtration is to remove dust, dirt, grinding dust, and any foreign object which would circulate in the hydraulic fluid. Should the oil start to oxidize, some products of oil oxidation will be removed. Normally, oil should not be allowed to degrade to that point.

Portable filter carts are available in many plants. Some maintenance programs use them to filter the oil as it comes from the drum into the hydraulic system. Most drummed hydraulic oil is very clean and would not contain silica or metals over 10 parts per million. Some oil distributors, who have authority to repackage bulk oils, have been known to refill a returned drum without having that drum reconditioned. This is an extremely bad practice which could introduce contamination into the hydraulic system.

The degree of filtration is dictated by the micron rating of the filter medium. It would be very difficult to filter too fine but the filter costs increase as the filtration size decreases. The finer filters cost more money and have a shorter life. If a machine has a 10-micron filter, the fluid will gradually become saturated with particles smaller than 10 microns and will have to be changed out.

Most filtration will not remove water or coolant contamination from a hydraulic fluid (more on these two contaminants under Troubleshooting). Filters are least efficient when they are first installed. As the filter medium begins to collect contaminants, the pores of the medium become slightly clogged and become more efficient. This action continues until the pores become so small the differential pressure across the filter becomes excessive and the filter begins to by-pass.

Although rare, pleated paper elements can become brittle and crack at the folds. Contaminants will then pass through the crack as the path of least resistance. Filtration problems can also occur with a cotton waste filter if the oil being filtered finds the path of least resistance and bypasses through the medium without being filtered. In either case, the hydraulic oil is being circulated through the filter but is not being filtered.

The cost-effectiveness of a filter is to use it for its full life and then replace it when the pressure differential is high or it has been in intermittent use for a year. If you should note the pressure steadily increasing and then dropping, that could mean the filter has become defective and should be changed immediately.

Reclamation

Reclamation of industrial hydraulic oils can be achieved easily if the volume in the plant warrants the expense. This section deals only with R&O or AW hydraulic oils. It is very difficult to reclaim engine oil used as hydraulic fluid because the additive packages in it will not allow many contaminants to settle out.

Certain contaminants cannot easily be removed. Chemical or water-soluble coolants cannot be effectively removed because of the water content and additives in the contaminating fluid. Hydraulic oils so contaminated can be used in miscellaneous or slop oil applications such as chain oiling, wire rope, and other noncritical applications.

Water and dirt are the two major contaminants which can be removed easily in a settling tank. The process can be accelerated by raising the temperature of the oil in the tank up to 170 to 180°F for several hours to reduce the viscosity of the oil. The reduced viscosity will allow solid contaminants and water to settle out more rapidly.

A vertical settling tank should be built on legs with a cone bottom where the water, grinding dust, and general dirt will collect. A valve on the bottom of the cone will allow the contaminants to be drained off the hydraulic fluid in the tank. After several days of settling and after the contaminants are drained off, the hydraulic fluid can be filtered out of the settling tank into clean used drums for makeup to the hydraulic systems.

This is cost-effective since most hydraulic fluids are contaminated before their oxidation life or ability to perform is impaired. In these days of accelerating costs and environmental considerations, reclamation can be an important step in cost avoidance.

If the hydraulic fluid has come from a machine which uses a corrosive cutting oil, be sure there is no cutting oil contamination in the hydraulic fluid. The sulfur and chlorine in the cutting oils can be corrosive to some nonferrous parts in the hydraulic system. If there is any doubt at all, use the reclaimed oil for noncritical applications as stated above.

Troubleshooting and Oil Sampling

Troubleshooting a problem hydraulic system starts by determining the problem. The most common problems encountered are these:

1. Short pump life
2. Slow hydraulic response time
3. Short oil life
4. Foaming of the oil in the reservoir
5. Soft mushy hydraulic response
6. Short filter life
7. Overheating of the hydraulic oil

Now that the problem has been identified, these are the operational factors which must be evaluated in an effort to determine the cause of the problem. These factors are the most common:

1. The proper viscosity and type of oil is not in use.
2. The correct oil level has not been maintained.
3. Scheduled maintenance has not been performed.
4. Filtration is not operating properly.
5. Hydraulic reservoir is not properly sized to the pump.
6. Contamination potential has not been minimized.
7. Pressure relief is not properly set.
8. Hydraulic reservoir temperature is excessive.

Should you find the suspected cause of the problem immediately, don't stop looking at the other potential causes of poor performance. It may be a combination of factors and you may be able to identify a problem which is just waiting to happen next week. Do a complete evaluation.

Take a small test tube sample of the oil circulating in the system. Make sure the sample is representative of the oil circulating in the hydraulic system. If the system has been down for several days, any contamination will have settled to the bottom of the reservoir.

If the system is down and has been cleaned out, you will be unable to get a sample which could assist you in finding a solution. Do not accept a sample from the drain pan used in draining the oil from the system. It is useless.

Allow the sample to settle in the test tube standing upright. Watch for dirt settling out, water which will show a clear break with the oil or coolant which will remain like a milky haze in or below the oil phase of the sample. If you have water and coolant contamination, you will see water at the bottom of the test tube, hazy coolant between the water and oil, and then oil in the top part of the tube.

No hydraulic fluid can protect a system with these contaminants circulating in the oil system! This contamination will contribute to short pump and filter life.

If there is a large amount of *dirt* in your sample, look for the following causes:

1. Dirt in the bottom of the reservoir that was never cleaned out during the last oil change.

2. A filter that is bypassing, allowing unfiltered oil to circulate in the system.

3. A filter system without some means of determining the pressure differential or a filter that has no pressure differential on the gauge. This indicates that there might be a hole in the filter, the bypass is stuck open, or it is a brand new filter. Inquire about the filter change program. Talk to the operator and maintenance person.

4. Check the reservoir fill cap and breather for damage that would permit dirt to enter. This might be the cause, depending on the conditions in the plant. At least, it will be a contributing factor.

If there is *water* in the sample, which will clearly break out from the oil phase of the sample, the hydraulic oil must be changed out. Look for the following causes:

1. A leaking oil-water heat exchanger or oil cooler.

2. How did they clean out the system when they changed the oil in the system? Was a water-based cleaner used and perhaps not totally cleaned out?

3. Check how the drums of new oil are stored and if water could enter the drum prior to the oil's being added to the system. Look for rust or rust rings on the top of the drum to determine if they have been wet. Oil drum bungs are *not* waterproof. If an oil drum is stored flat outside and it rains, water will collect on the top of the drum. The sun expands the oil and air in the drum during the day. At night, contraction takes place and the water is sucked into the oil drum. When the pump is inserted into the drum, it draws the oil-water mix right

off the bottom of the drum. Do not accept oil drums from your supplier with evidence that water has laid on top of the drum.

4. Any other possible source of water contamination.

If there is *coolant* in the oil, it will show as a cloudy phase below the oil but above any water present. Discuss with the maintenance people how the contamination can be reduced or stopped from occurring again. Coolant contaminated hydraulic oil must be changed out.

If the system is *sludged or varnished,* oxidized oil or contamination with a cutting oil is probably the cause. Check drain intervals and operating temperature to ensure they are correct. If a system is operating at a temperature of 170 or 180°F, well above the recommended range of 120 to 125°F, make sure the drain intervals have been shortened to compensate for the increased rate of oil oxidation.

Should you notice a slight burned odor to the oil, it could have been subject to high spot temperatures in the system. The odor is not necessarily an indication that the oil is bad. Oxidized oil will show an increase in viscosity compared with new oil.

If an adverse condition is causing problems in one machine, check other machines to ensure those conditions do not exist there. If they do exist, you are going to keep coming back to solve problems all the time unless those conditions are corrected or steps are taken to live with them.

Talk to the machine operators and the maintenance people who service the machines. They know what is really going on in the plant. Be very careful how this information is conveyed to your maintenance manager. You do not want to become involved with departmental politics.

Some hydraulic systems are designed to fit into a certain amount of space, and the reservoir may be undersized for the hydraulic pump. A good guideline is for the reservoir to be about three times the volume of the hydraulic pump rating. A 20 gallon per minute hydraulic pump should have a reservoir of at least 60 gallons. This allows the oil adequate dwell time in the reservoir for the oil to dissipate air, some heat, and some contaminants it might have picked up in circulation. A hydraulic system with a reservoir that is too small for the pump will show foam, high temperatures, and short pump life.

If the machine operator is experiencing slow or mushy hydraulic response of a machine, the cause is typically low oil level or air in the hydraulic oil. The low oil level is easily solved. Air in the oil might be due to that low oil level or a very small leak in the intake side of the hydraulic pump. As the pump picks up the hydraulic oil, it also sucks in some air and mixes it well on the way to the controls. Since air compresses, the result is mushy hydraulics.

Slow hydraulic response is normally caused by a hydraulic oil which is too viscous because the wrong product is in the system or the operating temperature is too low. Both are easily corrected. Some high-precision machine tools are very critical of the oil viscosity.

Short filter life can be caused by water or coolant contamination and, on occasion, oil oxidation. Rarely will metal particles be abundant enough to cause short filter life.

Overheating of the hydraulic fluid can show that the heat exchanger is defective or the water or air supply to the heat exchanger is restricted. Allowing the hydraulic system to bypass large quantities of oil will also increase the oil temperature.

If there is any doubt about the cause of the hydraulic problem, take a representative sample from the circulating system and send it to the laboratory for analysis. If water and coolant is evident in your test tube sample, a laboratory sample is a waste of money.

Properly maintained, a premium-grade hydraulic oil will provide several thousand hours of service in an industrial plant. The cost of purchasing and maintaining expensive machine tools demands a premium-grade lubricant to reduce machine downtime and increase productivity and profit.

Important: After you have completed your investigation, report back to the maintenance supervisor or manager and advise them of your findings and the probable causes of the problems in a memo form. Recommend the changes necessary to correct the problem. Compliment the maintenance team for the fine overall job well done and, if appropriate, thank them for the opportunity to assist in the hydraulic maintenance program.

Self-Quiz

Mark your answers to the questions and compare them with the answers listed at the end of the book.

1. Name three basic types of common hydraulic pumps.

    a. _____

    b. _____

    c. _____

2. If an oil is classified as "R&O," what additives does this represent?

   a. _____

   b. _____

3. What does the performance of any hydraulic fluid largely depend upon?
   a. Oxidation inhibitors
   b. Antiwear additives
   c. Proper maintenance

4. Automatic transmission fluid (ATF) is never used as an industrial hydraulic fluid because of its detergent additive level.
   a. True
   b. False

5. What would be the normal water content of a water-in-oil inverted emulsion fire-resistant hydraulic fluid?
   a. 37 percent
   b. 45 percent
   c. 50 percent

6. An "inverted" emulsion means that:
   a. Parts of oil are surrounded by parts of water.
   b. Parts of water are surrounded by parts of oil.

7. Fire-resistant hydraulic fluids are safe because they will not burn.
   a. True
   b. False

8. No special seals are required in hydraulic systems using any FR hydraulic fluid.
   a. True
   b. False

9. Many informed people feel this level of filtration is adequate for industrial hydraulic systems.
   a. 0.5 micron
   b. 8 microns
   c. 10 microns
   d. 30 microns

10. A paper element filter becomes less efficient during service up to the point when it must be changed.
   a. True
   b. False

11. If a hydraulic oil is contaminated by cutting oil, it can still be reclaimed by settling if it is returned to the same machine from which it was drained.

a. True

b. False

12. The bungs in an oil drum are designed to be waterproof and will prevent water from entering the drum.

    a. True

    b. False

13. AW hydraulic oils currently contain what additive to protect the pumps?

    a. Zinc compounds

    b. Sulfur-phosphorus additives

    c. Dyes and graphite

11
Industrial Gear Oils

Types of Gears

The types of gears commonly found in industry are the spur, bevel, helical, double helical or herringbone, and the worm gear. Hypoid gears, common to automotive use, are not common in industry.

The spur gear has teeth cut straight across the gear, perpendicular to the body of the gear. Since the gear teeth are cut straight across the gears, the gear shafts are parallel. Gear tooth contact is limited to a very few teeth. See Fig. 11-1.

The teeth on a bevel gear (see Fig. 11-2) are cut on a cone-shaped base, and the shafts are normally, but not necessarily, intersecting. The angle of cut determines the angle of the shafts. The teeth on spiral bevel gears are cut to increase the tooth contact surface over that of the spur gear.

A helical gear (Fig. 11-3) is the same as a spur gear but the gear teeth on both gears are cut at an angle, This increases the tooth surface contact of the teeth in mesh. A helical gear set develops thrust which tends to force the gears to opposite sides, and this thrust must be absorbed by the thrust bearings.

A double helical is two helical gear sets on the same shafts with the gear teeth cut in opposite directions. This arrangement tends to cancel out the thrust of each individual gear. If you were to take those two helical gears and combine the driving gears into one gear and the driven gears into one gear, it would be called a herringbone. It is illustrated in Fig. 11-4.

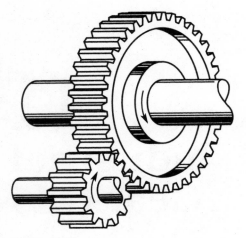

Figure 11-1

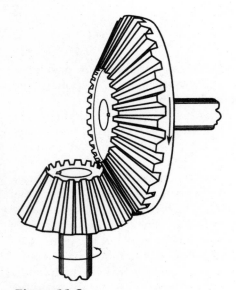

Figure 11-2

The last type of common industrial gear is the worm gear (Fig. 11-5). It is very common today even though it is the least efficient of the different gear sets. Properly cut worm gear sets are able to carry very heavy loads and provide high gear reduction ratios. The action between the worm and the gear is all sliding, rather than rolling, motion and re-

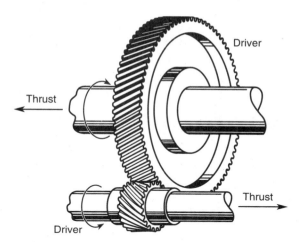

Figure 11-3

Figure 11-4

quires special lubrication consideration. Normally the driving gear is the worm, and the difficulty in lubrication depends on whether the worm is located on the top or bottom of the driven gear. If the worm is on the bottom, it can be bathed in oil, but the worm on top has to depend on oil carried up to it by the driven gear.

Figure 11-5

Types of Industrial Gear Oils

The basic types of gear oils for industrial applications are as follows:

1. Rust- and oxidation-inhibited mineral oils

2. Extreme-pressure gear oils

3. Steam cylinder oils

4. Automotive gear oils

5. Synthetic gear oils

6. Semifluid greases

7. Open gear lubricants

The American Gear Manufacturers' Association (AGMA) has set viscosity standards for gear oils starting with AGMA 1 to 8A for R&O oils, AGMA 7 Comp to 8A Comp for compounded gear oils, and AGMA 2EP to 8EP for extreme-pressure gear oils. These classifications do not apply to semifluid grease or open gear lubricants.

Viscosity

The viscosity needed to protect the gears is calculated by the design engineers and may range from an AGMA 1 up to an AGMA 8A Comp in viscosity. The viscosity recommendation will vary with the temperature, speed, type of gear, and method of application in or on the gears.

1. A gear set subject to *high* temperatures, *slow* speeds, or a *splash* lubrication system will require a *more* viscous oil.
2. A gear set subject to *lower* temperatures, *faster* speeds, or a *pressure* lubrication system will require a *less* viscous oil.

With a splash lubrication system in a gearbox, the oil adheres to a gear and is carried up and around the gears as they mesh. A viscous oil will cling to the gears better than a less viscous oil and is needed to protect the gears.

With a pressure circulating oil system, the oil is pumped up and sprayed on one or more points of mesh, ensuring an abundant supply of lubricant. Because the oil does not have to cling to the gear teeth, as with a splash system, a less viscous oil may be used. Should the pressure circulating oil system become disabled at any time and the gear set have to continue to operate with splash only, the gear oil should be changed to a more viscous oil.

R&O Mineral Oils

Quality mineral oils, formulated with rust and oxidation inhibitors, are the primary lubricant recommendation for precision gear sets like Falk and Farrel and do an excellent job. Extreme-pressure gear oils are recommended when the application of the gear set involves high pressures or shock loading.

Oxidation life and demulsibility are excellent with the R&O turbine oils in higher-temperature applications. A typical R&O oil recommendation by a gear manufacturer would be an AGMA 4 or 5 at normal ambient temperatures.

Without expressing an opinion about the quality, these major oil company trademarked products, and others, fall into this general type of product:

Shell Turbo Oil

Chevron Turbine Oil

Unocal Turbine Oil

Mobil DTE Oil

Texaco Regal Oil

Extreme-Pressure Gear Oils

This type of gear oil, probably containing rust and oxidation inhibitors, is fortified with an extreme-pressure additive package. The EP package is normally sulfur-phosphorus. Some companies have added fillers like moly and Paratac to enhance the performance of the basic EP package. The moly mechanically separates the gear teeth while the EP package will separate the teeth by chemically reacting with the metal to reduce the welding and tearing which could take place under heavy loads. The addition of moly or graphite will make the gear oil black in color. The Paratac additive makes the oil sticky and stringy but could shear down in service and lose its effectiveness.

Water demulsibility and oxidation life is fair to good but does not normally compare with a quality straight R&O oil. Some oil companies try to get away with using a less expensive stock oil in this type of product, and when they do, rapid oxidation and some foaming of the oil is the usual result. Naturally, they sell the product for a lower than normal price.

Without expressing an opinion about the quality, these major oil company trademarked products, and others, fall into this general type of product:

Shell Omala

Mobilgear 600 Unocal

Chevron NL Gear Compound

Texaco Meropa

Unocal Extra Duty NL Gear Lube

Steam Cylinder Oils

Steam cylinder oils are just that, viscous oils originally for steam cylinder applications. Since there is a severe lack of steam cylinders in industry today, the major use of this type of product is in worm-type gear sets and other applications requiring a slippery fatty additive. The normal product is approximately an SAE 140 (AGMA 7) or more and contains 5 to 6 percent fatty compounding.

In steam cylinder applications the viscosity assists in maintaining the oil film in hot applications and the fat is used to prevent water from washing the lubricant off the cylinder walls. Today, the viscosity and fat reduce metal-to-metal contact in applications with primarily sliding friction.

Some inexperienced oil salespeople try to avoid using this type of product in worm drives by recommending the EP gear oils in the recommended viscous grade. With the temperature rise of approximately 90°F over ambient, the EP additives can become chemically active and, in time, stain or corrode the nonferrous parts in the worm drive. Oxidation life should be better with the steam cylinder oil compared with EP gear oils. Discounting synthetics, compounded steam cylinder oils are still the best recommendation for this type of gear set.

Although compounded steam cylinder oils are not cheap, they are cost-effective in most applications. If the application requires a lot of effort to service and industrial injury exposure is high for the oiler, synthetics would be far more economical in the long run.

Without expressing an opinion about the quality, these major oil company trademarked products, and others, fall into this general type of product:

Texaco Cylinder Lube

Chevron Cylinder Oil

Unocal Steaval

Mobil 600W Cylinder Oil

Shell Valvata

Automotive Gear Oils

Although automotive gear oils are designed primarily for hypoid gearing, there is an occasional industrial use for them. Basically, they are similar to the industrial extreme-pressure gear oils in formulation but contain about double the EP additive package. They also contain a viscosity index improver to meet the MIL-L-2105D specification for automotive use.

Although you may see a manufacturer's recommendation for an SAE 90 gear oil, this does not mean it is an automotive gear oil. SAE 90 is also the industrial grade of ISO 220 or AGMA EP 5 grade in viscosity. Some equipment manufacturers use the SAE designation because so many oilers, used to SAE grades, can easily relate to it.

Additives other than the EP package and oxidization inhibitors are rarely if ever added to automotive gear oil by the manufacturers.

Without expressing an opinion about the quality, these major oil company trademarked products, and others, fall into this general type of product:

Unocal MP Gear Lube LS

Shell Spirax HD

Chevron Universal & Ultra Gear Lube

Mobilube HD

Texaco Multigear Lubricants EP

Synthetic Gear Oils

Some oil companies and blenders are starting to emphasize the use of synthetic gear lubricants in industry. The PAO (polyalphaolefin) synthetic stock is common and the performance is outstanding. These products are typically priced 300 to 400 percent higher than the mineral oil gear oils. In many applications, however, they are very cost-effective if the customer can mentally get by the price shock. An example would be an overhead conveyor gearbox where an oiler must climb a ladder to service it. You will risk an industrial injury and excessive labor cost for the oiler to get and climb a ladder several times to change out and refill that gearbox.

That gearbox may hold only one quart of oil. It would be much better to convert to a synthetic lubricant which would not have to be periodically changed.

Just the nature of the base fluid can be advantageous because of its inherently better oxidation life, lower internal friction, cost savings in increased gearbox efficiency, and lower power costs. Some major worm gear manufacturers are now factory filling their gearboxes with the synthetic.

Some of these products can pass EP tests without the use of EP additives and can give you extended drain intervals with complete safety. They are available with or without extreme-pressure additives in a variety of SAE and AGMA grades.

Because of the variety of base fluids, it is impractical to list any commercial products here.

Semifluid Greases

A semifluid lithium grease, in an NLGI 00 or 000 grade and containing a viscous oil and EP additives, can be a problem solver in a leaky gearbox. Some old slow-speed gear sets in heavy industry leak badly but are too valuable to shut down and/or expensive to rebuild. The seals are shot and the box leaks oil. The gears are usually badly worn and have large clearances.

Here are some examples where a semifluid grease, in a gearbox, can provide you with outstanding service and keep a valuable piece of equipment running until repair or replacement is possible.

1. On a mill runout table, the gear drives were in poor condition and leaked about one drum of oil per month per box. The gear drives were drained and an NLGI 000 lithium EP grease was installed. The semifluid grease clung to those gears like honey and the leakage was reduced to one drum per year per box.

2. On an overhead crane gearbox, leakage could not be controlled to prevent the oil from completely leaking out before the next service interval. The same 000 grease was installed and the problem was solved. The semifluid grease performed very well down to 0°.

3. All the grease had leaked out of another worm gear set because of the badly worn seals. Upon opening the gear set, it was noted that all the semifluid grease was gone except for that which was clinging to the gears themselves. That was enough to prevent a failure of the gear set.

The biggest disadvantage to the semifluid grease is the difficulty of draining it out at the service interval. It should be used only to control leakage in older and slow-moving gear sets or in units where leakage will be controlled at a later time by overhaul.

This can be a real problem solver and an area in which you can provide service and dollar savings for your company.

Open-Gear Lubricants

At best, open-gear lubrication is difficult. The gears could be exposed to airborne dirt and water and receive only periodic application of lubricant.

The most common of the open-gear lubricants is an asphaltic-based lubricant cut back with some type of diluent or solvent for ease of application. It can be applied by spray, brush, or paddle, and when the diluent evaporates, a tough black plasticlike coating is left on the gears.

The lubricant will be distributed from gear to gear as the gears go through mesh but only as long as some diluent remains. An open container of this type of product must be kept sealed to reduce the evaporation of the diluent.

Another type of open-gear lubricant is similar to the cut-back product except it does not contain the diluent. It is designed to be heated and sprayed on the gears hot. It will congeal more rapidly than the diluent type and will provide the same protection to the gear teeth. This type of product is more common in very heavy industrial applications such as steel mills or cement plants.

Some other products are actually greases with a very viscous oil and will leave a heavy greaselike coating on the gears. This type of product would be redistributed on the gears as the gears move through mesh. The disadvantage of the grease-type open-gear lubricant is that it squeezes out of the mesh area of the gear teeth more readily than the asphaltic-type product. Moly and graphite fillers, along with rust and extreme-pressure additives, would be added to the product in an effort to enhance performance.

Products like these are difficult to evaluate, one against another, since different manufacturers use different viscosity asphaltic bases. The application viscosity of the finished products is also different because of the type and percentages of diluent used.

Self-Quiz

Mark your answers to the questions and compare them with the answers listed at the end of the book.

1. Steam cylinder oils normally contain an extreme-pressure additive.
   a. True
   b. False
2. S.A.E. has established the viscosity ranges for all industrial gear oils.
   a. True
   b. False
3. Which type of industrial gear oil has the longest oxidation life?
   a. R&O oils
   b. Synthetics
   c. EP gear oils
4. Which type of additives would you choose to protect the gear teeth against shock loading?
   a. R&O additives
   b. Fatty compounding
   c. Extreme-pressure additives
5. Which type of additive would you choose for a worm gear application?
   a. R&O additives
   b. EP additives
   c. Fatty compounding

6. Grease should never be used in an enclosed gear set because it will tend to channel.
   a. True
   b. False
7. Synthetic gear oils require an EP additive package to protect the gears in hot applications.
   a. True
   b. False

12
Machine Tool Lubrication

Introduction

Machine tools come in all different types and designs but are made from simple components. Besides gears, bearings, and cylinders, we could be transmitting power by hydraulics or a transmission.

It is strongly recommended that you gain all the information possible from the equipment manufacturer's books or your oil supplier's books before making lubricant recommendations. The products currently being used may not be the products which are most correct. Many people, when asked if they have any problems, will answer negatively. To them, poor performance may be the normal so there is "no problem." Your job is to look for a better way to get the job done efficiently and cost-effectively in an effort to decrease machine downtime and increase profits for your company!

Some machines have a common reservoir for a multipurpose lubricant that will handle both the hydraulics and the ways. Other machines will have separate systems to handle each function. It is vital to learn the particular features of a machine tool to ensure that the correct lubricants will be used.

Hydraulics

Many machine tools operate with a hydraulic system. The most common hydraulic fluid recommendation would be for an R&O or antiwear hydraulic oil in an ISO 32 or 46 grade. Although many consumers tend to standardize on an antiwear ISO 46 grade, some complex machining centers require the ISO 32 grade to operate at maximum efficiency. The small increase in viscosity by going from an ISO 32 to 46 will slow the machine operation down to a snail's pace. This is unacceptable. There are exceptions to every rule, and in some machines, an ISO 22 spindle oil might be called for in some special hydraulic attachment to a machine tool.

Many machine tool hydraulic systems have oil coolers as mentioned in Chap. 10. They are very desirable for keeping the reservoir temperatures within the desired operating range of 120 to 130°F, but unfortunately some are also a common source of water contamination. In-line hydraulic filtration is common on machine tools, and servicing should be part of the routine maintenance program.

Some machine tools use a common system which needs an oil to perform both the hydraulic and way lubrication functions. Some manufacturers recommend an ISO 68 way oil and others recommend a special product which will act as a hydraulic fluid in noncritical applications and eliminate chatter and stick slip on moderately loaded ways. If your oil supplier does not have the recommended type of product, recommend your purchasing agent purchase an outside product that will meet the manufacturer's requirements. It is best to have the right product.

The hydraulic oil used in a machine tool is the lifeblood of the machine. People buy machine tools to produce profit, and if a machine is down it is not profitable. Quality hydraulic oil, along with proper maintenance, can keep that machine operating with a minimum of downtime. If the downtime is 10 percent or more, your plant has big problems. Check to be sure purchasing did not downgrade from a "quality" product to a "price" product in its quest for a lower price. Use quality, service, long life, and cost-effectiveness in your product as a guideline rather than just price.

Filtration, both in-line and portable, is vital to the proper operation of a machine tool hydraulic system. Control the temperature, keep the oil clean and dry, and it will provide outstanding performance in your hydraulic systems.

Without expressing an opinion on quality, these major oil company products, and others, fall into this general class of product:

	R&O oils	Antiwear
Shell	Turbo T	Tellus
Mobil	DTE Named	DTE 20
Texaco	Regal R&O	Rando HD
Chevron	Industrial Oil R&O	AW Hydraulic Oil
Unocal	Turbine oil	Unax AW

Spindles

Machine tool spindles are usually close tolerance antifriction bearings which operate at very high speeds. Because of these speeds and close tolerances, the bearings normally require a highly refined, rust- and oxidation-inhibited, low-viscosity oil. Some very high speed spindles require a spindle oil so light it has a viscosity similar to diesel fuel at room temperatures.

Spindle oils are available in a variety of grades ranging from ISO 2, 10, 15, to 22. The higher the speed and the closer the tolerances, the lighter the recommended viscosity of the oil. Back to the principle of "High speeds require light oils."

Without expressing an opinion on quality, these major oil company products, and others, fall into this general class of product:

Shell Tellus Oil

Texaco Spintex Oil

Sun Solnus Oil

Chevron Spindle Oil

Mobil Velocite Oil

Some Filmatic spindle bearings require an R&O oil in an ISO 32 grade. *Do not* use antiwear oils in these bearings.

Other spindle bearings are lubricated with grease, and the machine manufacturer's recommendation should be closely followed. The bearings may require a grease containing a very low viscosity oil for high speeds. NLGI grade 1 or 2 greases are most common. There is a potential for coolant contamination in some spindle bearings, so be alert to this problem.

Some spindle bearings and other bearings may be lubricated by mist systems. Here the oil is misted into the bearing area so just a very small amount of oil will "wet" the bearing surfaces. Too much oil will cause undesirable drag in the bearing. The viscosity of an oil which is misted follows the basic rule: "High speeds and light loads require a light oil—low speeds or high loads require a heavier oil." Some mist oils for heavily loaded bearings are required to be more viscous and may have to be heated prior to misting. Mist oils typically contain rust and oxidation inhibitors and some type of antiwear, film strength, or extreme-pressure additive. Follow the manufacturer's recommendation.

If a conventional R&O oil is used in a mist system, the oil will not readily drop out of the mist and could migrate into the air outside the machine, causing a health hazard. A well-formulated *mist* oil will easily reclassify into larger droplets and drop out of the air in the bearing area.

Way Lubrication

Ways are the parts of a machine tool which slide one against the other. They may be horizontal or vertical, flat, or V-shaped, but they will move one against the other. Way oils are specially formulated with a noncorrosive additive to provide anti-stick slip and a tacky additive to provide the required adhesive properties.

Stick slip is the result of moving one surface across another. It takes more energy to get a surface moving than it takes to keep it moving. This action will give you a jerky motion which, in a machine tool, will give an erratic motion between the workpiece and the cutting tool or wheel. This erratic motion will cause errors, defects, and rejected workpieces. Way oils are formulated to reduce this erratic motion, provide the proper oil film between the moving parts, and adhere to the surfaces of the ways. They may be applied by oil cup or an automatic system which meters the oil to the application or may bleed into the contact point from the combination hydraulic-way reservoir.

Without expressing an opinion on quality, these major oil company products, and others, fall into this general class of product:

Mobil Vactra Oils

Texaco Way Lubricant

Unocal Way Oil HD

Chevron Way Oil

Shell Tonna Oil

Grease Applications

A multipurpose lithium, extreme-pressure grease in an NLGI grade 1 or 2 is the most common grease in an industrial plant. There is a trend toward high-temperature thickeners, such as lithium complex, which are completely suitable for multipurpose use. Special applications, such as very hot or cold applications or centralized systems, may require additional greases to be stocked.

Troubleshooting

Most of your trouble with machine tools will be in the following categories:

1. Contamination
2. Extended drain intervals
3. Poor filtration
4. Improper lubricant application

These four categories exclude one of the most common problem areas and that is "no maintenance." Some small machine shops feel that, if there is still oil in it, maintenance is not that important. Even small companies must realize that profit comes from finished parts going out the door. Preventive maintenance keeps those finished parts going out the back door and the profits coming in the front door.

Contamination

In the hydraulic systems, watch for (1) water contamination from the oil coolers, (2) solvent contamination from cleaning the machines, (3) cutting oil contamination, and (4) spindle or way lubricant leakage into the hydraulic system.

In the lubrication of the ways, look for hydraulic oil leakage and/or cutting fluid contamination. Cross contamination will not destroy machines but will reduce the efficiency of the way lubricant and shorten component life.

Extended Drain Intervals and Poor Filtration

Watch for plugged or restricted intake screens in the oil reservoir and neglected hydraulic filters that are bypassing. Hydraulic oils must be kept clean and dry and must operate in the desired temperature range. Hydraulic oil which has oxidized will darken in color and take on a rancid odor. Continued use of oxidized oil will leave a coating of varnish on the interior parts of the hydraulic system and cause poor operation. Hydraulic filters should be changed on a regular basis to ensure cleanliness.

Portable filtration or a portable filter buggy can be used to clean up dirty systems or provide supplemental filtration to problem systems. Ten-micrometer filtration or better is fine for normal machine hydraulics. Numerically controlled or computer-controlled machine tools will benefit from filtration finer than just 10 micrometers. If there is water or contamination, do not bother to filter. Find the source of the contamination, correct it, and then change out the oil and the filter.

Improper Application

Sad to say, this is a common problem in machine shops. The oiler will use a product that is handy rather than walk the length of the plant to get the proper lubricant. Some maintenance people and machine operators still think "oil is oil." Take a sample of the problem lubricant, both new and used, and if there is an apparent problem, send it into a lab for analysis or use the service provided by your oil supplier. Set the drum which was the source of the problem aside and open up a new one until the matter is cleared up.

Supplier Services

About once a year, your supplier should hold an informal clinic for the maintenance personnel and the oilers. The clinic should consist of a complete listing of all petroleum products in the plant and a complete explanation of their formulation and uses. Continuing education on lubrication is well worth the time taken in training your personnel.

A smart supplier will arrange to have coffee or cokes and doughnuts along with some promotional "giveaways" for the people who dispense the products. If the plant is working three shifts, arrange a schedule with management to cover all the personnel even if you have to assist your supplier in holding a clinic at 2 in the morning. The properties of each lubricant should be explained to dispel the notion that oil is oil. Get small test tubes or clear bottles of product samples and show them to the group. Then the group can see, feel, smell, and even taste them. Explain that sometimes the color of a product might be slightly different because the color of the base stock oils might be different from batch to batch. Further explain that color is not a manufacturing specification for that reason. Tell them you need their assistance to make everything in maintenance work well. These educational clinics will help the mechanics and oilers to understand the types of products in the plant and how they differ in their applications.

Personal performance and quality maintenance is as much *perceived* as it is reality. It is important that the *perception* by the maintenance personnel become a reality.

Self-Quiz

Mark your answers to the questions and compare them with the answers listed at the end of the book.

1. A good hydraulic reservoir operating temperature would be:
   a. 120 to 130°F
   b. 130 to 140°F
   c. 140 to 150°F
2. If your industrial plant wants to standardize on one viscosity of hydraulic, the following ISO grade would be a good choice:
   a. ISO 46
   b. ISO 68
   c. ISO 32
3. High-speed spindle bearings require a heavier oil, like an ISO 46, to provide good lubrication and long life.
   a. True
   b. False
4. Some Filmatic spindle bearings require antiwear oils.
   a. True
   b. False
5. Mist oils can also be used in common hydraulic-way lubrication systems.
   a. True
   b. False
6. Grease is never used in a high-speed spindle bearing.
   a. True
   b. False
7. If your maintenance team says they have no problems, don't insult them by inquiring further.
   a. True
   b. False

13
Compressor Lubrication

History

Compressed air has been used in all types of industries for over 1000 years. References to it can even be found in the Bible which tells of compressed air being used in the smelting of silver, iron, and other metals. The use of compressed air came into its own during the industrial revolution when water power, steam engines, and then electric motors were used as the primary source of power. Today, compressors can be driven by any means of power including gasoline, diesel, and natural gas engines and electric motors.

Modern compressed air started out with a simple single-stage reciprocating compressor. The heat of compression became a problem so water was used to cool the cylinders either by injection or with water jackets. It was found that if the heat could be controlled, higher pressures could be attained using two- and three-stage compression with cooling between stages.

Other types of compressors, such as rotary, centrifugal, and axial-flow, were conceived and patented over 100 years ago but many years passed before practical machines could be produced. The first rotary was probably a simple blower used in smelting metals.

The two major problems with the compression of air are heat and moisture, and the chosen lubricants must be able to cool, lubricate, seal, and in many cases, resist the washing action of excessive moisture.

Compressor manufacturers clearly state what type of lubricant is required for their machines. It is important to follow their recommendations in the selection of the product. Their recommendations are normally based on petroleum products, but more and more of them are recommending and selling their own brand of synthetic lubricants for longer, trouble-free operation.

Moisture

Water vapor exists in the same space and at the same temperature as the air. The ability of the air to hold more water increases as the temperature rises and decreases as the air is cooled. In the high temperatures of the desert, the relative humidity is low, while in the cooler temperatures, the relative humidity is higher. As the air is cooled, it will reach a point where it can no longer hold water, and the water will start to condense. This temperature is called the "dew point."

When air is compressed, it takes up less space, but the space occupied by the water remains constant. Therefore, the relative humidity increases. Then as the compressed air temperature decreases, the relative humidity increases even more and will condense out of the compressed air. The cooling and containment of compressed air, even in the dry desert, will produce considerable moisture. Naturally, even more moisture would result in a humid climate.

For example, in a two-stage compressor taking in air at atmospheric pressure, 70°F, and 75 percent relative humidity and discharging it at 120 psi, over 3 gallons of water per hour will be condensed in the intercooler per 1000 cfm of free air. That is a lot of water! In simple shop air compressors several inches of water can accumulate in the bottom of the tank every day; it should be drained off to ensure low-moisture air.

Maintenance should be aware of this fact and schedule maintenance to eliminate the water from their compressed air coolers, air tanks, and air lines. If excessive water is allowed to accumulate, water will be blown down the air lines and into the compressed air applications like tires and pneumatic tools, causing corrosion, wear, and ultimate failure. Everything possible must be done to eliminate as much moisture from the compressed air as possible.

In some large compressed air systems and systems for critical equipment, some type of air-drying equipment should be used to ensure a dry air supply. Desiccant can also be used to supplement this equipment and supply the driest air possible to protect precision equipment.

Air-Line Oilers

Many air lines will incorporate a final water separator and in-line oilers in the air line to provide dry air with light lubrication to the compressed air application. If the air is relatively dry, a rust- and oxidation-inhibited ISO 32 mineral oil will do the job very nicely in air line oilers. If the application involves heavy duty air tools, a specially compounded air tool oil will be needed to protect the tool from extreme pressures and any moisture left in the air. A small hand tool would require a light-viscosity oil while a jack hammer would require a more viscous oil compounded to provide protection from extreme pressures. Some heavy-duty air tools have their own oil reservoir.

Reciprocating Compressors

Reciprocating compressors are like an engine with a source of power to move it rather than internal-combustion power. The combustion chamber is where the air is drawn in, compressed, and expelled. The reciprocating compressor has one or more stages and may be air- or water-cooled. The lubrication is the crankcase lubricant which would normally splash feed or pressure feed oil to the other components. The configuration may be vertical or horizontal and even might compress air on both sides of the piston where it is separated from the crankcase.

Regular scheduled maintenance of the valves is required in many configurations to ensure proper operation. An R&O ISO 100 or 150 mineral oil is a typical crankcase recommendation. If the cylinders are supplied by a separate lubricating system, different oils may be appropriate depending on the temperature and the degree of moisture in the air compressed. This recommendation varies from an ISO 68 up to and including an ISO 460. Teflon rings in the pistons make cylinder lubrication less necessary but longer life can be achieved with small amounts of lubricant. Synthetic fluids are becoming more common in "recip" compressors and are very cost-effective, providing long drain and valve maintenance intervals. Synthetics may also provide excellent protection with less cylinder lubricant used.

Rotary Compressors

Rotary, centrifugal, and axial-flow compressors all employ rotating elements in compression rather than reciprocating pistons. Lubrication varies with the different rotary compressors, but the principal parts to be lubricated are the timing gears, the shaft bearings, and in some cases, the thrust bearings and step-up driving gears.

The sliding surfaces in a rotary sliding vane compressor require lubrication to minimize friction and wear and to seal the small clearances between the vanes, rotors, cylinder walls, and head. This application normally requires an ISO 100, up to an ISO 320, turbine-quality mineral oil, depending on the temperatures involved in the application.

Centrifugal Compressors

These compressors are very good at supplying large volumes of air or gas at relatively small increases in pressure. At lower pressures, they could be called "fans." They may be lubricated with just grease cups to the bearings, grease-packed bearings, an oil slinging ring in an oil bath, or a pressure circulating oil system to main lubrication points.

Oil recommendations will vary with the manufacturer and service but an ISO 32 to 68 R&O turbine-quality oil would be a common recommendation.

Axial-Flow Compressors

These compressors resemble a steam turbine in some respects. Rows of fixed and moving blades alternate, and the flow is axial rather than radial. These units are high-speed, highly efficient machines for large volumes of air with compression ratios of up to about 12:1 or higher. They are typically powered by electric motors, steam turbines, or gas turbines. This kind of unit is used for very high volumes of air such as a test wind tunnel, blast furnace blowing, or underground mine ventilation.

Lubrication is similar to other high-speed rotating machines, such as centrifugal compressors and steam turbines. Operating temperatures are normally moderate in rotating-type compressors. The same principle: higher temperatures require heavier oils, lower temperatures require lighter oils, applies to this and all other applications. Bearing loads are normally moderate.

The correct lubricant must have good film strength, chemical stability, oxidation resistance, rust protection, and the ability to separate from condensation water.

The correct grease must have high oxidation resistance, mechanical stability, correct resistance to oil separation, and good antirust properties.

Synthetic products are becoming more important in compressor lubrication, with some compressor manufacturers marketing their own brands. Superior results can be obtained in some units with synthetic products, and they can be very cost-effective. The diester, fluorosilicone, polyglycol, and PAO synthetics are the types most commonly used. It is very difficult for maintenance management to change their mindset from mineral oils to synthetics but the long-term results would justify that change. A very important point to remember is that the synthetics are *not fire-resistant*. Special fire-resistant fluids must be used in some applications and where insurance or safety requirements dictate it.

Air Filtration

Huge volumes of air are taken into an air compressor, and air cleanliness is vital to a compressor's long life. A variety of air filters are in use today to remove dust, soot, and other fine particles of foreign material. The four main types of air intake filters are dry, viscous-impingement, oil-wetted, and oil bath.

Dry-Type Filters

The dry type utilizes pleated paper, like an automotive air filter, or special felted material formed in pockets protected and held by a wire mesh or similar material. The filtration surfaces are not oil-wetted, and they may be cleaned by a vacuum-type cleaner or compressed air; some may even be washed with soap and water and then dried. After any filter element has been cleaned, it must be inspected for holes or little cracks in the element to ensure continued efficient service. This is normally done by viewing the element from the outside with a light source on the inside. Any imperfections can be seen in the element. If imperfections are found, the element should be replaced with a new unit.

Viscous-Impingement Filters

The viscous-impingement type is a boxlike arrangement with metal mesh filters which are arranged in an increasing density toward the air outlet. The mesh is oiled with a light tacky oil before being placed into service. As the air passes through the filter medium, impurities are "impinged" on the oily mesh. This type of air filter can handle large vol-

umes of air and has the capacity for a large amount of impurities owing
to its overall surface area. The mesh can be cleaned with a solvent,
dried, reoiled, and put back into service.

Oil-Wetted Filters

This type of filter is basically like the viscous-impingement type with
oil-wetted mesh but arranged in a circular fashion similar to the dry-
type filter. It can be easily cleaned, dried, reoiled, and put back into
service. Again, a lightweight tacky oil should be used, and some products are marketed as filter oils. In an industry plant, a lightweight machine way oil may be used as a substitute if a more tacky oil is not
available.

Oil-Bath Air Filter

This type, in a smaller version, was used on automotive engines years
ago. In essence, it is a pan of oil covered by a canister and depends on
air velocity and change of direction to remove foreign particles. The air
enters at the upper outer edges of the canister, travels down through or
past the oil bath, reverses direction, and passes through an oil-wet metal
mesh for final cleaning. As the air changes direction, most of the larger
foreign particles are thrown into the oil bath and the remaining particles
are removed by the oil-wetted mesh.

The oil bath can be cleaned out easily and changed to eliminate the
captured dirt. The oil-wetted mesh can be removed and cleaned, if necessary, although the oil dripping down off the mesh will normally
carry most of the fine particles with it into the oil bath. This type of system performs best with a relatively large volume and a relatively constant volume of air. Its performance is substandard at low air velocities
such as at idle position. The viscosity of the oil chosen is dictated by the
ambient temperature of the air. An oil too light in viscosity may be
sucked out of the filter and into the compressor by the incoming air. An
oil too heavy would not drip off the mesh and carry the residual dirt
with it.

The cleanliness of the air cannot be overstressed, particularly in a
dusty, dirty environment. It is a maintenance item that can be easily
overlooked, and it is suggested that air filter maintenance be continuously stressed to the compressor maintenance team. Clean air is vital to
any operation.

Self-Quiz

Mark your answers to the questions and compare them with the answers listed at the end of the book.

1. List the four main types of air compressors.

   a. _____

   b. _____

   c. _____

   d. _____

2. Any moisture in the compression of the air can easily be handled by the use of a compounded lubricant.
   a. True
   b. False

3. If a jack hammer is being used on the end of the air line, a special high-viscosity oil must be used to protect the jack hammer.
   a. True
   b. False

4. The relative humidity of the compressed air will (increase or decrease) as it is cooled.
   a. Increase
   b. Decrease

5. The use of synthetic lubricants is not cost-effective in axial-flow or reciprocating compressors.
   a. True
   b. False

14
Metalworking Fluids

Overview

Metalworking fluids is an extremely broad field that would require a book by itself to fully cover the subject. The proper selection and use of a metalworking fluid depends on the machinery involved, metals being machined, speed and amount of the cuts, activity and viscosity of the proposed fluid, type of production runs, and other things in general.

It takes time and experience to properly select and use most metalworking fluids. A major manufacturing plant decided to transfer the cutting oil responsibilities from an old time industrial engineer to a couple of new engineering graduates. Upper management related that cutting fluids were not really a "black magic art" and the new engineers could gain valuable experience from the opportunity. Within four months, the results prompted management to return the responsibility to the engineer experienced in machining, metals, and cutting fluids. They found the combinations made the selection of cutting fluids very difficult. Many of the variables can be changed to achieve the desired production rate tool life and finish on the final part. In this sense, cutting fluids might easily be classified as a "black magic art."

There are so many types of cutting fluids for so many purposes that it would be impossible to list even a good representative sample. All the major oil companies carry a full line of products at reasonable prices, but their representatives may be lacking the experience to assist you in the proper selection. Although they do have metalworking specialists, your exposure to them will be very limited. This situation has given an excellent opportunity to small specialty companies who have specialized in selling and engineering metalworking fluids at higher profit margins than the major oil companies enjoy. Your need for this expertise will naturally vary with your operation.

Metal Machinability

Commercially available charts list various metals by their ease of machinability. The higher the machinability number the more easily that metal can be machined as compared with a standard steel such as 1020 steel. Aluminum alloy, magnesium alloy, and leaded bronze fall into the easy-to-machine category. Metals like copper, cast iron, and mild low carbon steel fall into the medium range while titanium, high-carbon steel, and stainless steel fall into the difficult-to-machine range. Typically, as the difficulty of machining increases, the more chemically active the cutting fluid should be to achieve a balance between part finish and tool life.

Machining Operations

Different machining operations add another difficulty factor to the differences in the machining difficulty for each metal. Operations such as turning, honing, drilling, and milling are easy compared with gear cutting, thread grinding, deep hole drilling, and gear hobbing which fall into the medium range. Those operations which are classified as difficult are tapping, broaching, and threading. As the severity of the machining operation increases, the chemical activity should be increased.

Some of the operations, such as broaching, will also require an increase in the viscosity of the cutting fluid because the fluid must be viscous enough to cling to the cutting tool. In other operations, such as deep hole drilling, where cooling, chip flushing, and penetration into the cutting area are important, a decrease in the fluid's viscosity may be necessary.

In addition, there is always the machine operator factor. Within a given set of metals and operations, a good machinist can make a marginal cutting oil perform or fail depending on the speed of the operation and tool setup. If you have been charged with setting up a cutting fluid test, obtain the best machinist you can find to assist you.

Some plants have machines set up to continuously produce the same parts from the same metals while other plants are continuously changing metals, parts manufactured, and the cutting fluids necessary to do that job. The plant which is continuously changing is more like a short-run job shop than a production-type operation. Requirements at each plant vary and there is no normal. This further complicates the selection and maintenance of the cutting fluids required.

Many times the correct fluid will be chosen to produce the best parts and the longest tool life for the majority of the parts manufactured. The use of this cutting fluid is then expanded to smaller operations where its performance is barely satisfactory in an effort to reduce the number of fluids in the plant. On short production runs, tool life is often immaterial because the tools will be changed at the end of the run anyway.

Cutting Fluids

The usual purposes for cutting fluids are to cool the workpiece and the tool, provide lubrication and chemical activity between the tool and the workpiece, and prevent welding of the worked metal to the tool.

Cooling is required to control the very high temperatures which will occur between the tool and the workpiece by friction. The cooling is required both to hold the size of the workpiece and for the life of the tool. Lubrication is required for good tool life as is the chemical activity of the cutting fluid. As the tool meets the workpiece and begins to remove metal, a plastic or almost molten part of the metal being removed will be formed behind the tip of the tool.

The chemical activity of the cutting fluid will control the formation and should allow the correct amount of the molten portion to develop. If there is too much antiweld or chemical activity, the molten portion will be too small and the tip or cutting portion of the tool will wear very quickly. If the cutting fluid has too little antiwear, the molten portion is too large and portions of it will slough off the front of the tool and degrade the finish of the work piece. Too much antiweld will shorten tool life and too little antiweld will affect the workpiece finish. The correct degree of antiweld, chemical activity, or antiwear in the cutting fluid will depend on the dragginess and shear strength of the metal being cut along with the cutting speed, finish required, and tool angles.

The cutting fluid additives which provide the antiweld activity include sulfur, fat, sulfurized fat, and chlorine. They may be present in any combination and in any amount. Highly sulfurized products are classified as corrosive. The use of chlorine in a cutting fluid may be prohibited if the finished part could be used in the aircraft or aerospace industry.

Mineral oil cutting fluid specifications normally show the degree of antiweld by listing the percentage of total sulfur, active sulfur, chlorine, and lubricity or fat. The difference between active sulfur, that which can be corrosive, and total sulfur is usually from sulfurized fat which is in the additive package. Unsulfurized fat can also be added for increased lubricity in some applications. Nonstaining or noncorrosive fluids normally contain no active sulfur, but the heavy-duty products in this class may contain some chlorine. Staining or corrosive cutting fluids can contain any of the antiweld additives in any combination and will show much higher copper strip corrosion test results in their specification.

There are two basic types of cutting fluids: mineral-oil-based and aqueous-based emulsions or solutions. Under each of the two types are the following basic types of cutting fluids:

Mineral Oils

1. Noncorrosive mineral oils without or with minimum antiwear. This type of product will normally be semitransparent, have a very mild odor, and contain sulfurized fat plus some chlorine in the heavy-duty grade. Their viscosities range from that of a spindle oil to an SAE 10 or 20. This is the type of product normally used in automatic screw machines.

2. Dual-purpose mineral oils for hydraulics and cutting. These are normally the noncorrosive mineral oils explained above. Since they are not really corrosive, they may be used as a hydraulic fluid, air-line oil, or spindle oil as well as a cutting fluid. In some machines where leakage and dilution of the cutting fluid is possible, this type of product in the hydraulic-spindle applications can be a major advantage in maintaining the cutting fluid's activity and efficiency. This type of cutting could be found in automatic screw machines or used as a lubricating oil in the machine heads or reservoir which would leak into the cutting oil reservoir. This reduces the contamination of the cutting oil.

3. Corrosive or active mineral oils containing sulfur, sulfurized fat, and/or chlorine. They are normally darker in color than the noncorrosive products, have a stronger odor, and will stain copper. The higher additive level is required for many metals and operations. They are also available in grades from a heavy spindle oil for applications such as deep hole drilling, up to an SAE 10 or 20 for applications which require the clingage of a more viscous oil.

4. Antimisting cutting oils. These are available in almost any one of the mineral oil cutting fluids. They are a feature of a cutting fluid rather than a separate fluid and contain an additive similar to that used in a misting lubricating oil. The purpose is not to eliminate misting of the oil but to allow it to rapidly drop out of the air within the machine area. In a congested machine shop where regular cutting fluids are used, it is not uncommon to see cutting fluid dripping from the ceiling. The use of an antimisting cutting fluid will greatly reduce this housekeeping mess, the potential for industrial injury, and the health hazard.

Aqueous Products

1. Water-soluble emulsions are a combination of oil, additives, emulsifiers, and water. The compounded oil is added to the water in concentrations anywhere from 10 to 1 up to 60 to 1 depending on whether lubricity or cooling is needed the most. Emulsions provide excellent cooling, as in a grinding operation, low cost as illustrated by the ratios of water to oil, and little or no odor or smoke during use. The emulsion will quickly drain away from the chips leaving fair rust protection. The emulsion formed is a tiny droplet of oil surrounded by tiny droplets of water. The color is from milky white to brown, with heavier-duty products tending to have a darker creamier appearance. If you are having trouble keeping the product in an emulsion form, check the hardness content of the water being used to form the emulsion. Some of the oil-in-water fluids contain extra emulsifiers to form better emulsions in hard water.

One big problem with these emulsions is the potential growth of bacteria in the machine. Although they will normally contain bactericides, they will deplete, and growth can start rapidly and increase with temperature. The emulsion should be changed regularly and/or treated with additional bactericides. If bacteria growth has started, it is best to drain, clean, and treat the emulsion sump so the new charge of emulsion is not a new breeding ground for the old bacteria. No foreign substance, such as unwanted food, should ever be thrown into the sump because it will act as more food for the bacteria. Circulation of the emulsion in the machine will reduce the bacterial growth. Tramp oil, if allowed to accumulate on the surface of the emulsion, will contribute to bacterial growth and should be removed on a regular basis.

Caution: A cutting fluid containing water must never be used to machine magnesium or any magnesium alloy since there can be a reaction between the magnesium and water, creating a fire hazard!

2. Synthetic solutions are becoming more common today in larger companies that machine various parts with various operations and metals. These products can be very versatile and cost-effective for general use. Many operators like them because they do not have an "oily" feel and seem cleaner. The correct dilution of the synthetic is critical because too rich a concentration may cause defatting of the operator's hands. Following the directions from the manufacturer is vital for proper use and efficiency.

The synthetics do not support bacterial growth like an oil-in-water emulsion and are therefore more easily maintained as well as more expensive. They contain antiweld additives and wetting agents which add in cooling in heavy-duty operations where efficient heat removal is necessary.

Both oil-in-water emulsions and synthetic solutions are excellent for operations which require rapid cooling. Grinding is a perfect example, where lubrication, cooling, rust protection, and rapid removal of the chips along with loose abrasive and bonding material from the grinding wheel are required. These products are also excellent in operations where the fast settling of the grinding dust and metal fines is advantageous.

Other Machine Methods

Electrochemical grinding, electrochemical machining, and EDM (electric discharge machining) are special applications which require individual consideration which is best supplied by the equipment manufacturers. Regular cutting fluids are not used in these applications.

Reclamation

Various methods can be used to reclaim mineral oil cutting fluids. The easiest is simple settling in either the machine's sump or a separate container. That separate container can be a 55-gallon drum or a specially built settling tank designed for reclaiming. With ferrous metals, magnetic separators are very effective in the basic removal of fines at the machine. A centrifugal separator or centrifuge may be used to separate the metal fines, with the clean fluid being transferred to a separate holding tank for reuse.

In any reclamation system, the first step is naturally the removal of the coarse fines by the use of a chip breaker which will remove the cutting fluid from the chip pile and then a chip wringer which will remove the fluid from the smaller chips and metal fines. The same results may be obtained more slowly by just allowing the fluid to slowly drain from the chip pile. The remaining chips and fines may then be collected and recycled or sold. The resulting fluid may then be transferred into settling tanks, perhaps heated to eliminate any water or soluble contamination and to reduce the viscosity of the fluid for faster settling, and the fine expelled from the bottom of the tank. A final step may be the pumping of the fluid through a filtering system to ensure very clean reclaimed fluid which can then be reused. Additives are replenished by makeup.

Reclamation of oil-in-water soluble cutting fluids is normally not cost-effective. It is much simpler and cheaper to break the emulsion, treat the water phase for disposal down the sewer, and sell off the waste oil. Because of their longer life potential, reclamation of the synthetics, by the above methods, is easy and cost-effective. The cost-effectiveness of selling clean dry chips and recycling the cutting fluid is good. A good point to remember is to check the concentration of any aqueous emulsion or solution to ensure that normal evaporation of the water phase has not increased the concentration to an undesirable level. Should that occur, the addition of water only will correct the ratio.

Skin Disorders or Dermatitis

These problems can occur with any worker in any industry. Anyone can react to anything at any time, but our concern is a worker's possible reaction to cutting fluids. Soluble oils and synthetics can defat the skin, and the alkaline nature of some additives can be a problem, as well as bacteria growth in the emulsions. Prolonged exposure to the mineral oil cutting fluids has also been known to cause a problem with some people.

Most problems, such as dry and cracking skin, pimples, and boils, are caused by poor hygiene and/or lack of the proper precautions at the workplace. Clothing that is not laundered frequently and lack of protective creams on the hands and arms can ultimately produce reactions in some individuals. These precautions must be stressed by the active participation and continuous education of the workers by management. The reaction of one worker does not constitute a valid enough reason to change cutting fluids which are performing well. Investigate other possible causes for the problem, such as clean work clothes, carrying an oil-soaked rag in the pocket, good personal hygiene, and the proper application, fluid cleanliness, and concentrations of the cutting fluid before jumping to other conclusions.

Summary

Because of the vast variety in metals and machine operations, we regret there are no hard-and-fast rules to use in the selection of cutting oils. Several general points are:

1. If the antiweld is too high, poor tool life could result.
2. If the antiweld is too low, poor part finish could result.
3. Never use a water-based emulsion or solution with magnesium.
4. Water only will normally bring an emulsion back to the proper dilution ratio.

Self-Quiz

Mark your answers to the questions and compare them with the answers at the end of the book.

1. A cutting fluid, which contains sulfurized fat, is always corrosive.
   a. True
   b. False
2. Sulfur in a cutting fluid is normally not used in finished parts for the aerospace industry.
   a. True
   b. False
3. No cutting fluid should ever be put into a machine's hydraulic system.
   a. True
   b. False
4. Water-based coolants are required on magnesium for maximum cooling.
   a. True
   b. False
5. Should a water-oil emulsion become contaminated with bacteria, it should be treated with bactericides rather than changed out.
   a. True
   b. False

15
Miscellaneous Lubrication

Wire Rope

Wire rope is commonly made up of strands of wire wound into a bundle, with that bundle being wound around other bundles of wire strands. Many are lubricated as they are manufactured. Wire rope will typically "wear" from the outside in but will "break" from the inside out. As the rope is flexed, the inner strands move against each other. Without lubrication, those strands ultimately break inside. As the rope is moved over sheaves, the outside layers take the abrasive wear from the sheaves and dirt from the environment.

The key to complete wire rope lubrication is to (1) keep the inside strands lubricated, (2) prevent abrasive wear on the outside, and (3) protect the strands of the wire rope from rust and corrosion. Unfortunately, that is easier said than done. An ideal lubricant for wire rope would be one that would penetrate the strands to the core yet leave a tough, nontacky layer of lubricant on the outside. The ideal place to apply a lubricant is the point where the wire rope bends over a sheave so some penetration of the lubricant into the strands can be obtained.

Some maintenance people in the mining industry maintain that a heavy, tacky lubricant on the outside only attracts abrasive dirt. The tacky film plus the collected dirt shortens the life of the wire rope. Still others contend that the tacky film on the outside of the wire rope reduces the abrasive wear and is a good tradeoff against the dirt it may collect. One way of protecting the wire rope is to oil it as it bends with a wiper just following the point of application. An application of hot oil would allow more penetration of the lubricant and, as it cooled, would provide the necessary viscosity to protect the strands. A lubricant, cut back with some solvent, would also allow for deeper penetration. The wiper will remove the oil which would cling to the outside of the strands but allow some oil to remain and penetrate the inner strands.

The method used must be compatible with the working conditions in the plant or of the rope. Obviously, you cannot have a sticky open-gear lubricant on the rope if it is to be handled by the workers or can easily pick up abrasive material. On the other hand, a wire rope which is dripping oil inside a plant could create a safety hazard for the workers in the area.

Wire rope manufacturers typically prelube the wire rope as it is made, but the service life can be greatly extended by the proper lubrication. Application of the lubricant for your service requirements will vary. There is no one answer.

Chains

The key to properly lubricating chains is to get the lubricant into the links to lubricate those parts which move against each other. Chain drives can suffer far more harm from inadequate lubrication than by years of service. Many different types of chains are in service today and all require adequate lubrication in the link and pin connections to protect the components. The exception to this would be chains having wear-resistant materials in the linkage and operating in a highly abrasive atmosphere where a lubricant would only act as a grinding compound.

Slow-moving chains driven by small horsepower motors can be lubricated by a squirt can, oiled brush, or oil mist system. Medium-speed changes, those moving at 1000 to 2000 fpm, could be lubricated with a drip feed or sight feed oiler. High-speed chains may have a dip pan as part of their enclosure where others may be oiled by a sight feed lubricator with wicks extending to the linkage.

High-speed chains may also be pressure-sprayed or oil-misted. To assist in getting the lubricant into the linkage, consider drilling small holes in the links to allow the lubricant to enter. This is an excellent way to lubricate overhead chains and still minimize dripping of the lubricant on the floor below. Lubricant should be applied into the clearance between the links or to an oil groove, if present, on one side of each link.

For maximum chain life, it is important the oil source be kept clean and well maintained. Chain elongation and, hence, sprocket wear is greatly reduced with proper lubrication.

Lighter-viscosity lubricants are preferred for maximum penetration in the more precision chains while the heavier viscosities are preferred on the less precision chains.

Electric Motors

In most industrial plants, electricians handle the lubrication of electric motors. Many electric motor bearings, when serviced by plant maintenance people, fail from overlubrication because they give the bearings a "shot" when they are servicing the adjacent equipment.

Many electric motor manufacturers recommend a non-extreme-pressure grease because they feel the extreme-pressure additives can find their way into the motor windings and be detrimental. The recommendations vary from a light-viscosity synthetic fluid in the grease for lower temperatures or very high speeds up to a high-temperature grease with a more viscous oil to provide the proper oil film at the elevated temperatures. An NLGI 2 grade is the most commonly used grease and preferably contains no extreme-pressure additives or fillers such as graphite or moly.

Oil-lubricated motor bearings typically use a turbine oil in an ISO grade 68 for normal temperatures in a plant. Should the motor be outside in the winter, an ISO 32 type oil or a low-pour-point ISO 32 oil would probably be recommended. Under extreme conditions, such as very low or very high temperatures, a synthetic fluid would give excellent performance. Using our principles of high speed—low viscosity and low speed—high viscosity, low-speed motors would require a heavier-viscosity oil than would a high-speed motor.

Hand or Squirt Can Oiling

In some industries many applications are serviced with an oil can. Most are noncritical applications, so little concern is given to the correct oil. These applications are also normally once-through, meaning the oil is applied and may stay in the point requiring lubrication for only a short time or may gradually leak out. An oil cup would provide better lubrication if replaced by a spring-loaded oil dispenser. With a nipple attached, it could be installed at the point of application. This method would ensure a continuous supply of fresh lubricant into the application. All the oiler would have to do then is to add oil to the oil cup as needed.

Some plants stock an inexpensive squirt can oil, but you would be better off, if possible, to save your used hydraulic oil, way oil, or similar oil in a drum to use in those noncritical applications. The waste oil could be put into a drum, allowed to settle, and the cleaner oil taken from the top of the drum. Use of this waste oil would save you money and the disposal costs which might occur with the waste oil. The exception is squirt can oiling of electric motors. Only new fresh quality oil should be used to lubricate the bearings of an electric motor.

Water Contamination in Large Bearings and Gearboxes

A little trick which works in large bearings or gearboxes which are subject to water wash, as in a rolling mill or similar application, is simple air pressure. Most bearing seals are in place to prevent the lubricant from leaking out, but they may be susceptible to water entering the bearing cavity from the outside. One large steel mill found that by running an air line to the bearing and regulating the air pressure to a couple of pounds per square inch, the internal pressure in the bearing was enough to keep the water out. The same thing can be done with gearboxes which are subject to water contamination. Remember to keep the air pressure into the bearing or box at just a couple of pounds per square inch to create a positive pressure in the bearing or gearbox. It is a simple little trick that can save thousands of dollars in maintenance and protect expensive equipment.

Rust Protection and Equipment in Storage

Many kinds of rust-protection products are on the market. Basically, the length of protection is conversely equal to the ease of removal. The lightest-duty product is an interoperation rust preventive which is designed to protect machined parts between machining operations. The parts are typically dipped into the rust preventive, drained, and set aside. These types of products are moisture-displacing and will protect the machine part for a week or two of inside storage. The parts do not have to be clean prior to further machining. The interoperational rust preventives normally do not even have to be removed prior to the next machining operation.

Medium-term rust preservatives typically leave a waxy film and are meant for medium-length inside storage. Some are cut back with a solvent-type fluid for ease of spraying. They may be removed with a solvent prior to the part's going into service or becoming part of the finished product. The product may be applied by dipping, spraying, or even brushing the part to be protected with the rust preventive. Another type of medium storage product could be a volatile carrier for the rust-preventive additives which are polar in nature. This type of product will leave a dry or semidry film on the part rather than an oily or waxy film.

Rust-preventive additives can also be added to a conventional motor oil for engine storage. The oil is installed and the engine is run up to operating temperature and then shut down and sealed against air circulation. This method allows some of the rust-preventive additives to vaporize and protect the parts of the engine not protected by the oil bath. The engine may also be run for short periods of time without losing much of the rust protection. The same can be true for any type of circulating oil which could be used in a hydraulic system or gearbox.

Long-term outside storage products are, for the most part, never removed. They would be used to coat the base or anchor for a guideline where the anchor is to be buried in the ground. Long-term-type products can normally only be removed with very vigorous rubbing with a solvent. Should a machine be treated with this type of product, it is extremely difficult to remove prior to being put into service. Many of these types of products have an asphalt-base plus rust inhibitors and are cut back with a solvent-type fluid for ease of application by dipping, rolling, or being painted on the surface.

A gearbox or hydraulic system that has been internally treated with a rust-protection product can cause the initial fill of lubricant to foam. If that is the case, the initial fill of lubricant must be changed out, flushed with a cleaner, and then recharged with the recommended lubricant. This practice may have to be repeated until the foaming condition is eliminated.

Any component of a new facility which must be stored outside during construction should be protected and covered. Unfortunately, there is no "right" product to do the job. You should not use the long-term storage product with an asphaltic base because of the difficulty in removal. An interoperational light-duty product would not protect the component from the elements even though it may be covered. That leaves the medium-duty product which would leave a light waxy film on the outside of the component. If properly covered and provided with some ventilation, this type of product would probably provide the best protection and reasonable labor hours to remove the rust preventive.

16
Synthetic Lubricants

History

It is important to realize that the word "synthetic" is generic in nature just as is "beer." Work started on synthetic lubricants back in the 1930s and has steadily progressed to date. It has been said that Germany tried using them in their war machines to replace the conventional lubricants after the Allies destroyed much of the available petroleum refining capacity. There are many different types and many different uses for synthetic lubricants or hydraulic-type fluids. They fall into these general ASTM classifications:

Synthesized Hydrocarbons:

  Olefin oligomers

  Alkylated aromatics

  Cycloaliphatics

Organic Esters:

  Dibasic acid esters

  Polyol esters

  Polyesters

Other Fluids:

  Polyglycols

  Phosphate esters

  Silicates

  Silicones

  Polyphenyl esters

  Fluorocarbons

Blends.   Any mixture of the above. Some of these products may also contain some additives to enhance performance and mineral oil in some products.

Synthetic lubricants will, in many cases, provide superior performance in many applications such as extremely high or low temperature operation, compressor applications, fire-resistant fluids, and jet aircraft engines. Benefits may include longer service life, greater protection from oil degradation, fire resistance, reduced friction, lower operating temperatures, lower pour points, natural detergency, broader operating range, and a naturally higher viscosity index. Others in the synthetic classification offer water miscibility and fire resistance, such as the phosphate esters.

Some applications of the synthetic fluids are motor oil, automotive and industrial gear oil, aviation engine oil, aviation hydraulic fluids, fire-resistant fluids, compressor lubricants, industrial hydraulic fluids, and greases. Specifications for several military lubricants can be met only by a synthetic product. All commercial and military jet aircraft engines use synthetic lubricants, in addition to the space shuttle, NASA, and the atomic submarines.

In some applications the use of synthetic fluids is not practical: in circulating systems that leak oil, in systems that are exposed to extremely dirty conditions where the dirt cannot be prevented from entering the system, and in an engine that is so badly worn that oil consumption cannot be controlled. In these cases, the use of a long-lived synthetic is not cost-effective.

When conventional mineral oil lubricants are refined, the result is a complex mixture of different-sized molecules which also contain impurities, such as wax, silica, and sulfur compounds. These impurities cannot be completely and economically refined out of a conventional mineral oil lubricant. On the other hand, when a synthetic is synthesized, a consistent molecular structure can be achieved without the impurities.

Picture the molecular structure of a conventional mineral oil with layers of molecules, all different sizes, rolling up and down and bumping into other molecules. Now picture a synthetic fluid, with even, consistent molecules, rolling against each other. It is logical that the even, consistent molecular structure would have less internal friction between the molecules. This physical characteristic of the synthetic is the basis for the claims of less or reduced friction in an application. It also reduces the power needed to do the job. This can increase the miles per gallon in an automotive engine or increase the efficiency of a gear drive. Normally, the greatest efficiency gain will be in the applications which are the least efficient to start with.

Less friction will also mean less heat generated, which supports claims that applications with synthetic fluids will run cooler. Although claims vary, a temperature drop of 25°F or more is not unrealistic or uncommon.

Products

Polyalphaolefins

One of the most common of the synthesized hydrocarbon fluids (SHF) is polyalphaolefins, or PAO for short, which falls under the alkylated aromatic group of synthesized hydrocarbons. Ethyl Corporation, the leading producer of PAO stock in the United States, expects the worldwide demand to exceed production capacity by 1998 if planned expansion becomes a reality and demand curves move as expected. It is amazing that the growing synthetics lubricant market will comprise only 1 to 2 percent of the total worldwide lubricant market.

The PAO stocks are the most common base fluid for the products in the synthetic motor and automotive gear oil market today. Normally, a small amount of diester is added to adjust the seal swell rate, so seals commonly found in automotive engines will neither swell nor shrink. The additive packages are normally mixed with a small amount of mineral oil to get them into solution and then the mixture is added to the base fluid. On many containers, in small print, you will note that a product is 100 percent synthetic except for a small amount of additive carrier.

The synthetic motor oils are capable of superior performance for the following reasons:

Benefits Gained	Reasons for the gain
Extended drain intervals	Greater resistance to oxidation
Increase in fuel economy	Lower internal friction due to even molecular structure
Reduced wear in the engine	Great film strength of the synthetic as compared with mineral oils
A cooler-running engine	Less friction causes less heat. This relates to oil temperature and not the water temperature which is controlled
Better "crankability" in	Again, less internal friction added to cold weather naturally low pour points of the synthetic
Faster oil flow on start-up	Low pour points and less internal friction
Cleaner engines	The PAO stock has a high natural detergency plus the detergent-dispersant additive used

Many average motorists find the price of a quart of synthetic motor oil very high compared with the cost of a quart of conventional motor oil on sale at a discount department or parts store. Yet many who have switched to synthetic motor oil find that the entire cost of the oil change to synthetic can be paid for in the savings in fuel over the oil drain period. In addition, if proper oil drain intervals are observed, the engine will be free from sludge or varnish well past 100,000 miles.

Motorists who park outside in a cold climate will enjoy the advantages of rapid starting in the morning. Motorists who are stuck in traffic on a very hot day will enjoy the added protection for their engine and extended oil life that come with a synthetic. Those willing to spend a couple of extra bucks will find a synthetic motor oil and gear oil to be cost-effective in both performance and the longevity of their engine. It is, however, difficult to change the mind set about motor oil costing $1 per quart.

Car manufacturers, particularly General Motors, are now finding that a synthetic provides the only way some of their newest and hottest engines can live past the break-in stage. These engines are so stressed that conventional motor oils cannot adequately protect the engines. Therefore, they are factory filling with a synthetic motor oil and recommending its use to protect the engines.

One of the major advantages of the synthetic PAO stock is its compatibility with conventional mineral oils. Any addition of conventional oil for the synthetic product circulating in the system will, of course, reduce some of the superior qualities of the straight synthetic fluid. Any engine or circulating oil system can be converted to a PAO product without a major concern for contamination from the fluid used in the past. Naturally, it is advantageous to install the PAO product in a clean, dry system. If a system has been flushed out with a cleaner, the best route is to completely clean out the residual cleaner-oil mixture to ensure minimum or no contamination with the cleaning solution. If this is impossible or impractical, it is best to install a new charge of conventional product, run that charge for a day or so, and then drain and recharge with the synthetic fluid. Every step taken to reduce contamination between the cleaner and the synthetic charge will be cost-effective in the increased life and performance of the synthetic fluid.

PAO stocks were used in the first greases meeting the military specification MIL-G-81322C designed for aircraft grease applications. It was my understanding that the original tests were performed in the wheel bearings of an F-4 Phantom jet fighter. The F-4 would start down the runway and after reaching takeoff speed would slam on its brakes and abort the takeoff. With conventional greases, the rollers in the F-4 bearings would have been flattened out, but with the synthetic fluid, the airplane would be able to turn around and take off normally. Many industrial greases available today using different synthetic fluids are used very successfully in severe-temperature applications.

Semi- or Parasynthetics

Most of the "semi" or "para" synthetics on the market today are a blend of the PAO stock with a conventional product. The percentage of the PAO stock will vary and the superior performance characteristics of the synthetic will diminish as more of the conventional fluid is added to the PAO stock to form the finished product. Although some of the properties of the PAO stock will be imparted to the finished product, exactly how much is unknown and would depend on the percentage of the PAO stock. That is the main reason some people are very reluctant to use the semisynthetic products, and rightfully so. If you are increasing the cost of your lubricant by going to synthetic fluids, it is better to go all the way to 100 percent synthetic and gain all the advantages of the fluid. That is the only way to effectively compare the synthetic with the conventional lubricant. One advantage of the synthetic is a greater resistance to oxidation and, hence, longer service life. The addition of a conventional lubricant, as in a semisynthetic, adds an element that would oxidize at a normal rate. Since oxidation begets oxidation, the oxidation of the conventional portion could increase the oxidation rate of the synthetic portion.

Newer Semisynthetics

Some new "semisynthetics" on the market today are a blend of the newer superhydrotreated mineral oil stocks which have many of the superior properties of the straight PAO stock. These superhydrotreated stocks can be the base fluid and can be combined with PAO stocks or silicone to form a finished product which will supply superior performance as compared with conventional lubricants at a cost less than that of the straight PAO synthetic products. Their cost is only slightly less than that of a PAO stock, but that cost may be coming down in the future. It will never, however, be reduced to be equal to the cost of a conventional motor oil stock. It is possible that these newer superhydrotreated stocks will become the base fluid for a supergrade of conventional motor oils in the not too distant future.

The basic properties of the new semisynthetics blends may allow a notable reduction in feed rates for applications such as compressor cylinders. Reductions in the feed rates up to 90 percent have been documented vs. conventional mineral oils. This reduction can produce benefits which include less lubricant consumption, less lubricant going down the line, less maintenance in coolers and separators, and overall lower cost of operation and maintenance.

One major problem with this reduction in feed rate is the method of applying the minute amounts accurately and dependably to the compressor cylinders. Some of the older, and even the new, distributor systems which are designed for larger qualities of lubricant have this problem. Some systems are available, however, which are designed to handle very small amounts and have minimum internal leakage to ensure the accurate dispensing of very small amounts of lubricant. Although the dispensing system and lubricant will cost more initially, those costs will pay off rapidly to ultimately provide good cost savings. This method of lubrication can also prove cost-effective for those compressors with "no lube" rings by extending the life of these expensive Teflon ring modifications in compressor applications.

Fire-Resistant Fluids

The phosphate ester fluids are king in this field. They are used to replace conventional mineral oil lubricants and hydraulic fluids where there would be a fire hazard should an oil circulating hose break and spray flammable lubricant into a source of ignition. In many cases, a machine is built with special seals and designed for the use of phosphate ester fluids. The information from the manufacturer of your machine should indicate these special features.

To convert a machine not designed for a phosphate ester fluid is a major project. The machine will probably have to be dismantled and all seals will have to be replaced with those compatible with the fire-resistant (FR) fluid. If the reservoir is painted, the paint will have to be removed to the bare metal unless special paint has been used. Phosphate ester will soften and remove many paints. Most people will consider this an unwarranted expense and use another type of FR fluid, such as glycol and water or a polyol ester fluid, which are covered later in this chapter.

It is vital to remember that fire-resistant fluids are just that; fire-*resistant* and not fire-*proof*. The phosphate ester fluid will give a worker a chance to evacuate the area until the fire fighting force arrives or other measures are taken. It has been said that the fumes from a phosphate ester product will produce a caustic or poisonous gas when exposed to a potential source of ignition.

For those who need to use a FR fluid and cannot bear the expense of converting a machine to the phosphate ester fluids, other choices are available. One is the inverted emulsion type, which combines oil and water, and another is the glycol and water type. Both fluids contain about 45 percent water which will turn into steam when exposed to a source of ignition. The steam will snuff out the flames and allow a worker to safely retreat from the area. After the water phase has vaporized off, the residue of these fluids will still burn and the fumes can be hazardous.

Organic Esters

Dibasic Acid Esters.   Under the general heading of organic esters, probably the most common are dibasic acid esters, more commonly called diesters. A diester is the reaction obtained when combining a dibasic acid with an alcohol. Diesters have been used in automotive engine oils in the past, but most manufacturers have now switched to the PAO stocks for better performance. With a PAO-based synthetic motor oil, a small amount of diester is normally added to provide the correct amount of seal control with seals normally compatible with mineral oils.

Diester fluids are very detergent and can be used for cleaning a reciprocating air compressor of varnish and for solubilizing sludge left in the compressor by extended or poor performance of a mineral oil. They also make very good lubricants for the crankcases and cylinders of a reciprocating air compressor. Their use can reduce wear and deposits and their low volatility can reduce vaporization compared with a mineral oil.

Blends of diester fluids and mineral oils or superhydrotreated semisynthetic fluids are also available where a middle ground on cost might be a factor. Their primary use would be in reciprocating compressors.

Polyol Esters. Polyol esters are also excellent as hydraulic fluids and cylinder lubrication for high-pressure reciprocating and rotary screw compressors where cooling is generally inadequate. They have been around for a while but still perform well in many applications which are too severe for mineral oil application.

Polyol esters are also used as fire-resistant fluids in some hydraulic applications and compressors to reduce the chance for fires. Other major uses include jet engine lubricants which are used in most jet engines in commercial, private, and military jet aircraft. They are a major force in the total synthetic marketplace.

Other Fluids

Polyglycols. Polyglycols were among the first synthetic lubricants and are among the least expensive and most commonly used synthetic fluids. They are also known as polyalkylene glycols. They have excellent properties compared with conventional mineral oils and are commonly used in compressors, heat-transfer systems, refrigeration compressors, and as a bearing and gear lubricant in severe-service applications. They have very good performance in screw compressors handling LPG, natural gas, methane, and similar gases—even when the crankcase might be filled with that gas. This type of synthetic fluid is the type used, mixed with water, as a glycol-water fire-resistant hydraulic fluid.

They were also used as automotive hydraulic brake fluid back in the 1920s. In Europe, they are used in worm gears, highly loaded gear oil systems, and some refrigeration compressors. A wide range of molecular weight could be made, and with the proper selection of weights, water miscibility or immiscibility could be achieved. Unfortunately, one venture into a polyglycol-based motor oil was quickly dropped because the viscosity grades needed for motor oil were miscible with water and caused many problems in engines with condensation.

The disadvantages of polyglycol fluids are poor compatibility with mineral oils, the water miscibility factor in some applications, and the fact that they will soften or remove some paints.

Phosphate Esters.   This group of the organophosphate family has a phosphate radical incorporated into a hydrocarbon molecule, which provides fire-resistant characteristics to the fluid. The initial work on phosphate esters was done in the early 1940s to develop a fire-resistant aircraft hydraulic fluid to overcome the hazards of the hydraulic fluid coming in contact with hot brakes, exhaust manifolds, and other sources of ignition on an airplane should a hydraulic hose rupture.

Around the same time, an industrial version started to be marketed by a major company in the United States which was followed by other companies getting into the market in the 1950s and 1960s. Despite their excellent oxidation resistance, the problem encountered with these fluids was that when oxidation did occur, phosphoric acid and other by-products were formed. Naturally, the by-products led to corrosion, wear, and deposits in the system. Other problems were paint removal and seal compatibility. In general, epoxy paints and fluororubber seals are strongly recommended to combat these problems.

Also grouped under the "Other" category are silicates, silicones, polyphenyl esters, and fluorocarbons. These are regarded as specialty synthetics, are normally quite expensive, and are appropriate only for special and limited applications.

That covers the general line of synthetic-type products. They can replace almost every type of conventional mineral oil on the market today, but the cost-effectiveness depends on the conditions of the application. In many cases, they will be necessary to meet the increasing number of severe operating conditions of the machinery in industry today or their superior properties can prove to be more cost-effective than will conventional lubricants.

Their cost-effectiveness involves longer oxidation life allowing for extended service, reduced friction for a proven savings in power consumed to run an application, and the ability to operate effectively in extremes of hot and cold. Studies have shown it costs far more money to apply, maintain, and change out a lubricant than just the cost of the lubricant.

The big question is "Why has the world not embraced synthetics as the savior of the lubrication world?" The answer is simply "cost," and when viewed from a purchasing only standpoint, that is an obstacle. A PAO product is in a cost range of about $15 per gallon while a conventional lubricant is in the price range of about $3 per gallon. Imagine a maintenance superintendent telling the purchasing agent that they should buy a lubricant that costs 5 *times* what the regular oil costs. It would be very difficult for that purchasing agent to justify paying that much for an "oil." To justify the purchase, the maintenance manager will have to have figures which prove that the synthetic is "cost-effective," not just more expensive.

Many of the cost factors necessary to prove your case are listed in Chap. 1. They include the cost of changing out and/or cleaning out a system, machine downtime, fire hazards, extremes of temperature, savings of handling less product in the plant due to extended life of the lubricant, and others that would be applicable to your own operation.

Before approaching the purchasing agent, the maintenance manager should have calculated all these cost factors to determine the real cost of the synthetic lubricant or fluid. When presented with the cost savings involved from using the synthetic product, the purchasing agent should be able to see that the synthetic would be the cost-effective way to go.

Self-Quiz

Mark your answers to the questions and compare them with the answers at the end of the book.

1. The use of any synthetic will require special seals to be installed.
   a. True
   b. False
2. Which synthetic fluid is compatible with conventional mineral oils?
   a. Alkylated aromatics
   b. Dibase esters
   c. Phosphate esters
3. The only non-fire-resistant fluids in the synthetics listed below are
   a. Polyglycols
   b. PAO fluids
   c. Phosphate esters
4. Phosphate esters are the most common synthetic fluids used in automotive motor oils today.
   a. True
   b. False
5. The first fire-resistant aviation hydraulic fluid was not developed until the 1960s.
   a. True
   b. False
6. Dibase esters were used to make the first automotive hydraulic brake fluid in the 1920s.

   a. True
   b. False
7. Alkylated aromatics (PAO) fluids are compatible with mineral oils.
   a. True
   b. False
8. It is very unusual for a synthetic fluid to be used in a grease.
   a. True
   b. False

17
General
Recommendations

This chapter covers the general applications of lubricants in industry, mining, and construction. These suggestions are not set in concrete and should be altered by the recommendations from each machinery manufacturer. They are, however, designed to assist you in the consolidation of products, the selection of multipurpose products, what you can use in an emergency, and reuse of those products which have served their original function.

General Industry
Hydraulics

Standardize your hydraulic requirements with an antiwear, premium hydraulic oil in an ISO 46 grade. Other than special and critical machine tools which may require an ISO 32 grade to function properly, an ISO 46 grade is your best overall bet. An exception might be a system where the reservoir is sized too small, is covered with dirt, and runs hot all the time. In that case, you should correct the problems stated or use an antiwear ISO 68 grade hydraulic oil.

If you are ever accidentally out of stock of an oil, say an ISO 46, you may make one to carry you over until the proper oil is delivered. To roughly get an ISO 46, take *one* part of an ISO 32 oil and *two* parts of an ISO 68 oil and blend them together to make do. Do not mix a gear oil and hydraulic oil to use as a hydraulic oil because of the additional EP additives in the gear oil. An ISO 68 way oil may be used also.

If you are involved with some outside machinery in a colder climate, you may use a "low-pour" VI improved hydraulic oil in the same ISO grade. This type of oil will only help you on start-up. It is important in starting a really cold hydraulic system that the machine be allowed to idle and warm up prior to operation. It will harm the hydraulic pump if it cannot suck up a viscous hydraulic when under a load. A synthetic hydraulic oil could also be used in that cold hydraulic system.

The easiest way to check the condition of a used hydraulic oil is to take a rounded-bottom test tube sample from an operating system and allow it to sit upright for several hours. If the sample is foaming, you will note that as you take the sample. Look for the following conditions in the test tube:

Conditions	Probable cause
Dirt in the bottom	Poor filtration, an open fill cap, etc.
Dark brown color	Oxidation is increasing—suggest a change
Burned odor but good color	Not a big concern—oil might have been exposed to a small hot spot in the system
Burned odor and dark color	Change out—this indicates the oil is coming to or has reached the end of its service life
Clear moisture in bottom	Probably a leak in the water-oil heat exchanger or condensation if those conditions exist
Murky or an emulsion in bottom	Contamination from a way oil or metalworking fluid. Change it out
Gray slime in bottom	Has never really been defined, but start additional filtration or change the filter in the system and check again in a few days

Special Note: Should you ever have the occasion to add a defoamant to a hydraulic system, never, but *never* add more than the amount recommended by your oil supplier! To do so can create a condition called "air entrainment," which is tiny air bubbles in the oil that cannot be removed.

A premium hydraulic oil which is kept clean and dry and is operated at reasonable temperatures will have a long service life. Service life is also extended by excessive leakage of the hydraulic connections, and the system requires a lot of makeup oil. Every time you add oil, you are "sweetening" the system. This is not, however, an acceptable alternative to proper maintenance and drain intervals. It pays to fix the leaks rather than keep adding oil. A simple *one* drop every 10 seconds leak will cost you over 40 gallons of oil in a year. A *one* drop every second leak will cost you over 400 gallons per year plus cleanup maintenance costs. It is more cost-effective to fix the leak.

If a hydraulic oil has served you well and has reached the end of its service life, simple reclaiming by the settling method will allow you to reuse it in once-through noncritical applications, like a squirt can, wire rope, or a slow-moving chain lubricated by a semiopen oil pan. In some heavy industries that can still burn some waste oil and that burn No. 5 or 6 heavy fuel, the waste hydraulic oil can be added to the heavy fuel to be burned. It will add cheap Btus in the furnace, eliminate the disposal problem, and even make the heavy fuel move more easily down the pipe. The used hydraulic oil is probably cleaner than the heavy fuel.

Way Oils

There are basically four different grades of way oils available: ISO 32, 68, 150, and 220. Only two of them really make up the market, the ISO grades 68 and 220. The additives are the same and the ISO 68 is used in small faster machines while the ISO 220 is used in larger, slower-moving machines where the horizontal or vertical ways are more heavily loaded. You can also use the ISO 68 way oil as a hydraulic oil if necessary or recommended. The reason it may be recommended is that the manufacturer is fearful of hydraulic oil leakage into the way oil system which would reduce its efficiency in lubricating the ways. It is not worthwhile trying to consolidate way oils since only the two grades are in common use.

Way oils are also very good for general squirt can lubrication and, if possible, should be reclaimed and used for that purpose. Usage might include chains, noncritical bearings, hinges, or other applications which need a tacky oil.

Spindle Oils

Spindle oils, like way oils, typically come in four grades: an ISO 2, 10, 15, and 22. Probably the most commonly used is the ISO 10 but in the very high speed spindles, an ISO 2 is used. Remember, the faster the speed, the lighter the oil? There is no substitute for the proper spindle oil, so consolidation is not practical here. The low volume is rarely worth trying to reclaim and the use of reclaimed spindle oil in a spindle is not recommended—particularly at the prices for rebuilding the spindles.

Air Compressors

As covered in the compressor chapter, there are many different types of air compressors and there are many good compressor oils on the market. With the risk of fires, oil accumulation in the coolers, and high valve maintenance, the best lubricant that can be used is usually a synthetic one. Products are available which should be able to satisfy the requirements of the different types of air compressors, and you should be able to reduce your inventory of compressor lubricants to one or two products. Many feel the benefits of a synthetic compressor lubricant far outweigh the additional cost of the synthetic. The semisynthetics are a great improvement over conventional oils, but the synthetics really shine in this type of application. The majority, if not all, of the private branded compressor lubricants from the OEM are synthetics.

Greases

You can very easily standardize on a lithium complex thickened grease in an NLGI grade EP1 or EP2 which will handle 95 percent plus of your applications. Several years ago, the recommendation would be for a lithium EP grease, but the lithium complex thickener has all the advantages plus a higher dropping point for better high-temperature applications. There might be an application or two requiring a special grease, but that can be purchased in tubes to handle such small applications.

Don't get loaded down with a bunch of different thickened greases which may not be compatible with each other. The addition of one grease to another, in that case, could cause rapid softening or thickening of the combination. Greases with different thickeners are basically to be considered incompatible.

There is a special grease which can be a real lifesaver in special applications. That is a lithium thickened semifluid grease, in either an NLGI grade EP 00 or an EP 000. It is useful in gear cases, as will be covered later, and in plain bearings which will leak out and create a maintenance mess. The semifluid grease can reduce consumption, reduce leakage, and provide better protection for the bearing than the recommended oil. The machinery OEM will not recommend it, but some manufacturing plants have used it to their advantage. The grease must be applied with a spring-loaded grease cup or a distribution system designed to apply a given amount on a time cycle.

An exception in greases would be a grease for the lubrication of electric motor bearings. See Chap. 15 for that information.

One major manufacturing plant was experiencing short life with greasing flexible shafts. The plant happened to have some synthetic grease, meeting MIL-G-81322D, and found the grease with the synthetic fluid lasted 3 times longer than the regular shop grease.

Forklifts

There are still some propane or LPG forklifts running around plants today if they have not been replaced by electric units. On the engine, follow the OEM's recommendations for grade and viscosity. Synthetic motor oils have been found to last about 5 to 10 times longer than conventional motor oils in this type of engine.

You can use your industrial ISO 46 antiwear hydraulic oil in the hydraulic system without concern unless we are talking about outside cold conditions where a low-temperature hydraulic would be better in the long run. This should not be a problem. There are some OEM who will recommend automatic transmission fluid in the hydraulics, but it is not necessary unless cold conditions will prevail. Your antiwear hydraulic oil has the same antiwear agents as does the ATF.

For general forklift greasing and wheel bearings, use the lithium complex EP 2 grease. The transmission fluid should be the one recommended by the OEM. If it calls for a non-extreme-pressure SAE 90 gear oil, you can use any non-extreme-pressure ISO 220 oil in the plant or even an SAE 50 motor oil. If the call is for an HD or GL-5 gear oil, it is for an automotive extreme-pressure gear oil which should be used because of the high level of EP additives in it.

Cranes

A common problem with overhead cranes is gearboxes that leak oil. Your choices seem to be to (1) fix the leaks, (2) fill it to the proper level with the semifluid grease mentioned above, or (3) replace the gearbox.

A large steel mill was having trouble keeping a lubricant in the semiopen gear sets below a walking-beam furnace. Water leakage from the bottom of the furnace would fill the gearboxes and the oil would float on top until it overflowed, displacing the oil entirely. When a semifluid grease was tried, it was found that the grease adhesion to itself and to the gear teeth prevented the water from washing it all out of the gearbox and some boxes were actually operating under water.

One problem with using the semifluid grease in this type of application is that it is very hard to clean out when you want to change it. That supports the "no free lunch" theory. Other grease points on overhead cranes are easily satisfied with the lithium complex EP 2 grease in the plant.

Mist Lubrication

This is a rarely discussed application. A mist lubricant has to do two things; (1) mist easily into the area requiring the lubrication and then (2) fall out of mist so it does not mist out into the work area. It has been said that a major manufacturer in Milwaukee, Wis., was working with its oil supplier to solve the second half of the problem when it tried a 10W-30 motor oil and found it came out of the mist phase very nicely. It was deduced that the addition of a viscosity index improver was the additive needed. A VI improver was added to the oil, formulated to satisfy the other requirements, and mist oils were born.

It does not really matter if the story is true or not, but it does illustrate the properties required in a mist oil. Tiny droplets of oil must be carried by the air to the bearing or other point of applications and hit the bearing which requires the lubricating film of oil. Mist oils should not be subject to substitution or consolidation.

Construction

Motor oils are the most common construction site lubricants. They are used in engines, some transmissions, and hydraulics. The type used, however, will vary.

Most contractors have a mixed engine fleet which would include some gasoline engines, Caterpillar diesels, Cummins diesels, and Detroit Diesels, and all the OEM want different types of oil in their engines. The contractor is not going to stock a variety of engine oils to keep everyone happy.

Most contractors, depending on the climate, will select an API CD II or CE oil in an SAE 30 or 40 grade for general engine oil use. Some others might select an API CE oil in an SAE 15W-40 grade, particularly if the weather is cold. Diesel engines on a construction site are normally run hard and lugging is not a problem as it could be in an over-the-road truck. Coupled with good maintenance and recommended oil-filter changes, heavy construction equipment normally has good longevity.

For hydraulic applications, a contractor will normally use any SAE 10W or 20W motor oil for a couple of different reasons. First, that is the normal equipment's recommendation, and second, if it should accidentally be added to an engine as motor oil, it would not cause a failure. Problems can be encountered if a piece of equipment operates several hydraulic pumps off one hydraulic reservoir. A lot of heat is generated, and with marginal cooling and high temperatures, the temperature of the hydraulic fluid can be elevated over the recommended maximum. In that case, a contractor should go from the normal SAE 10W or 20W up to an SAE 30 to protect pumps from accelerated wear.

The most common gear oil is an automotive GL-5 grade, SAE 80W-90 extreme-pressure gear oil. In hot climates as in the southwest, an SAE 85W-140 is commonly used despite the OEM recommendation of SAE 80W-90. In hotter climates, contractors want the added protection of the heavier gear oil grade although it is rarely needed. The additional drag of the heavier gear oil is immaterial considering the power put through the gears.

Although some OEM and contractors recommend and use a lithium grease with up to 3 percent moly (molybdenum disulfide), most contractors stay with the straight lithium or lithium complex extreme-pressure grease in an NLGI EP 2 grade. The addition of moly to a grease will assist in the lubrication of sliding or oscillating motion, but it does very little for a rotating motion. It would therefore be beneficial in the lubrication of bucket pins which are rough plain-type bearing surfaces on a front end loader. This application must be greased after mere hours of operation, but contractors prefer to regrease every shift or during the lunch break if possible. A moly grease would assist in achieving this goal. The problem is that moly grease costs more than a grease without moly and contractors are very cost-conscious. They have competitively bid on a job, and additional costs and expenses can easily eliminate their profit.

Motor oil, gear oil, and grease are the main lubricants on a construction site.

Aboveground or Open-Pit Mining

Aboveground (AG) mining primarily involves automotive products because the majority of the consumption of lubricants is in the motor, hydraulic, grease, and gear oils for the haul trucks. Most larger mine sites handle the majority of their lubricants in bulk tanks and many do the major portion of dispensing those lubricants at a lube bay or building.

Twenty years ago, typical haul trucks carried about 75 to 85 tons. They kept getting larger and more expensive each year until their capacity is 200 tons or more and cost well over a million dollars. They vary from a diesel electric with the engine generating power for the individual wheel motors to diesel mechanical drive units using normal transmissions and differentials. Driving a large unit is like driving a two-story apartment building suspended on marshmallows.

The primary engines in these larger units are Detroit Diesel, Cummins, and Caterpillar and the horsepower is in the area of 2000. Naturally these engine are all turbocharged.

The Detroit Diesel 149 two-cycle engines are very common in large diesel electric haul trucks and their primary oil recommendation has been upgraded to an API CE, but they have run into some ash deposit problems so some mines have switched back to an API CD II oil with less ash. The recommended viscosity is an SAE 40 but if the temperatures are very hot, an SAE 50 may be used until the nighttime temperatures drop below about 55°F. This is not a major problem because many of these units run around the clock in a large mine.

Crankcase capacity will vary but will be around 50 gallons and have a consumption of about 2 quarts per hour of operation.

Cummins turbocharged four-cycle engines are also common, but Cummins prefers the use of an API CE oil with an SAE grade of 15W-40. Mines who predominantly use Cummins engines will normally use the 15W-40 in order to maintain the warranty.

The primary oil recommendation for the larger Caterpillar engines is an API CE and an SAE 30 grade. Some consideration is being given to changing that to a 15W-40 in the future. Although the Detroit Diesel and Cummins dominate in this market, the larger Caterpillar engines are making their presence known after a period of absence.

Underground Mining

Underground miners are a breed all to themselves. They go underground in the morning, work hard, tease and joke with each other, and come up at the end of the shift with a smile on their faces. The equipment they work with is different from that in any other type of mine. There are different types of mines and the equipment will vary with the type of mine, although there are mines with diesel equipment and their lubricant requirements are the same as those in an aboveground mine. The unique equipment is electric in mines that tend to have an explosive gas potential. Hydraulics are usually a fire-resistant water-oil combination, either an emulsion or an inverted emulsion to reduce the fire hazard.

An automotive GL-5 gear lubricant is the normal, but some mines use an NLGI 000 semifluid grease to reduce the leakage from gearboxes. Many mines are dark and are rarely illuminated by more than hard hat lamps on each person, so leakage from a machine is very difficult to detect.

Motors, reels, and miscellaneous applications are normally grease with an NLGI 2 EP grease and general oil applications are satisfied with an R&O mineral oil. Each mine will vary with the conditions at that mine, so there are no set rules.

18

Storage and Handling

Bulk Lubricants

Receiving lubricants in bulk is the most economical way to purchase them. Years ago the only way to receive bulk oil was with a large 10,000-gallon tank, either above- or underground. A secondary method was to receive and store products in small 500- to 1000-gallon aboveground tanks, but this method was restricted to locations that were close to the source of product by the manufacturer. With oil distributors becoming a more dominant force in the lubricant distribution system, the availability of small tankwagon (less than a transport delivery of approximately 7000 gallons) deliveries has become an important method of delivery to many industrial consumers, truck fleets, and car dealers. With the advent of smaller portable tanks, which hold around 500 gallons, flexibility has been further increased for the consumer.

The small portable tanks and bins are either round or square with channels in the bottom of them to accommodate forklifts and they can easily be moved from the storage yard to the points of applications. As the small tanks or bins get empty, they are returned to the lubricant supplier and are refilled. One advantage of the bin is that when it is returned for refilling, the new product is put into the tank on top of the product which remains in the bin and the customer is charged only for the product added. Whereas a 55-gallon drum would be returned with about 2 gallons of oil left in it, 100 percent of the product purchased in the bin is usable rather than wasted when the empty container is returned to the supplier.

Typically, the small tanks and bins were on loan from the supplying lubricant distributor. Although the product pricing might be the same as or slightly lower than drum products, there is a savings to the consumer in handling storage and labor in handling the drums which is not incurred with the bulk containers. The pricing is about the same as for drums because of the extra handling by the lubricant distributor. To enjoy this savings in in-plant handling, a forklift of adequate capacity must be available to handle the heavy load of the bin. Normally, breather vents are permanently installed in the top of the bin so the air which replaces the product removed is clean.

Bins are also useful in transferring product from a large bulk tank to points of application in the plant and are then refilled in the consumer's plant for further distribution. Periodically, the bins should be removed from service and cleaned, inside and out, to ensure the cleanliness of the product. When the bin is removed from service, vent filters and any cover seals should be renewed before being put back into service. The expense of this will depend upon your agreement with the lubricants and bin supplier. Some like the bins for different products painted a different color for ease of identification within the plant to reduce the possibility of misapplications.

Large bulk tanks, both above- and underground, must be properly vented to handle the venting of air when the tank is refilled and to properly filter the incoming air when product is removed from the tank. Regardless of all the filtration and care taken to ensure the product is clean in the tank, the tank must be completely emptied and cleaned out every several years to ensure cleanliness of the product. Pumps on large bulk tanks should be of low-pressure high-volume type to reduce the time taken to refill the smaller containers in the plant. Filters and meters can easily be added to further clean the product and monitor the amount distributed.

Not many consumers are large enough or have a large enough grease consumption to warrant buying bulk grease. It normally takes a bulk tank that will hold approximately 60 to 80,000 pounds for deliveries of approximately 45,000 pounds. Large steel mills and large aboveground mines might be able to take advantage of this product in bulk because they may have a few areas where large volumes of grease can be dispensed close to the bulk tank. Bulk grease tanks are normally aboveground vertical tanks and many outside are painted black to take advantage of solar heating of the grease. In cold climates, inside tanks will allow the pumping of the grease during the colder months. Naturally, the tank must be properly vented to handle the addition and the removal of product from the tank and to ensure the cleanliness of the air replacing the product removed.

If possible, for bulk grease tanks and bins, a screen should be installed after the valve to eliminate any large foreign particles which may have been in the tank or in the delivery truck. As in oil, everything possible should be done to ensure cleanliness of the grease applied to an application. Everything possible is done in the distribution system to ensure the delivery of clean product, and everything possible should be done at the consumer's location to further protect the product from contamination.

Drummed Product

Outside Storage.   Drums of product, stored outside, should be covered if possible. This can be done with a covered open-air shed, plastic sheeting, or drum covers. Plastic sheeting tends to be a nuisance because it must be removed and then replaced after drums are distributed and drum covers, which are just plastic or metal caps over the drum, tend to sail away in a windstorm. Drums are normally stored on pallets and should be rotated so the bungs are parallel with the bungs on the drum next to it. This way a 2-by-4 piece of wood can be inserted under the bottom edge of the drum, parallel with the bungs to prevent water from collecting around the bungs. Bungs on oil drums are *not* waterproof. If the drum is allowed to remain flat on the ground and it rains, the water will collect on top of the drum. As the sun warms the product in the drum, it will expand, and then it will cool and contract in the evening and suck the water right past the bungs and into the product.

Storing drums on their side is a good way to prevent water from entering the drum but one-third to one-half of the drums will leak product from the bung unless you first remove the seal and tighten the bungs down tight. Drums stored on their sides should rest on wooden rails to keep the drum off the ground.

Inside Storage.   Drums stored inside a building may be stored flat or in stacked racks, whichever is available in the plant. They can then be easily moved with a drum hand truck or forklift individually or by the pallet full to the place of application.

Two kinds of drum valves are inserted into the small or large bung openings of the drum. The size will be determined by the normal amount removed from the drum. The drum is then elevated so the distribution container can be inserted below the valve. Although many consumers just loosen the vent bung to allow air to replace the product removed, a filter vent should be inserted into the vent bung to ensure that filtered air replaces the product removed. Again, cleanliness is very important to machine longevity.

Grease drums come with a crimped-on solid top which will normally be removed and replaced with a grease pump and cover. It is important that the crimped cover of the new drum be removed prior to the removal of the grease pump from an empty grease drum. This will preclude the maintenance person from having to lay the pump on the floor prior to inserting it into the new drum of grease. This is just a simple procedure to ensure the pump does not get dirty prior to its insertion into the new grease.

Kegs, Pails, and Buckets. The next smaller container is a 120-pound keg, or a quarter drum, as it is sometimes called. They are normally used for grease or automotive gear oil. Although this is a commonly used size for smaller applications, it is also very good for grease applications which involve mobility of the container. The next size container, and a fairly common one, is a 5-gallon pail or bucket of oil or a 35- or 38-pound container of grease. The oil has a crimped-on top with a pop-up pour spout for easy transfer or application. A grease keg or pail has a solid crimped-on lid which will be removed and a grease pump and cover installed. The pump is then used to dispense the product into grease guns or into the point of application. Containers smaller than a pail are uncommon in an industrial plant, construction site, or mine.

The price of product in a pail will normally be the same as the drum price plus the cost of the container, which will be in the area of $5 per pail. This is pretty fair pricing when it is beneficial to the customer to use a small amount of a special product or when product must be transported by hand to the point of application. The trend is toward plastic pails today which can be cleaned out and used for other purposes in the plant so the additional cost is not all wasted. Pails of cutting oils are very handy to have for cutting fluid operations where the machine operator is responsible for keeping the cutting fluid level up during the day.

Special Container Sizes. With bulk, drums, kegs, and pails covered, the industry rarely needs any other products in small sizes, but there are exceptions:

Relatively small requirements for forklift motor oil or some small air compressors can be handled very nicely with a case of quart bottles. Motor oil is easy to handle and dispense in this size container.

Special greases for use in specialized applications can be made available in 14.5-oz cartridges which come 10 tubes per carton or 60 tubes to a master case. Cartridge grease is also very convenient for a maintenance person to carry in a back pocket to refill a grease gun as they proceed on their maintenance route if refilling the grease gun from a pail is impractical. When climbing on top of an overhead crane to grease bearings, carrying a spare tube or two is the most efficient way to carry more grease.

There are, of course, exceptions to every container and product, but this information should provide the basics of storage and handling of petroleum products. It is very important to stress that the main purposes of the proper storage and handling of lubricants are as follows:

1. Cost-effective purchasing of large-volume lubricants.
2. Ensuring that the lubricant being applied to the point of application is clean and free of contamination.
3. Choosing the proper container size for controlling labor costs for special applications.
4. Maintaining the proper amount of lubricant on the property to avoid costly shutdown of machinery due to lack of the proper lubricant.
5. Maximizing production and profits through proper lubricant handling.

Self-Quiz

1. Drum bungs are waterproof top openings on an oil drum.
   a. True
   b. False
2. Drums should never be stored upright in outside storage.
   a. True
   b. False
3. The smallest size container for grease is a 35- or 38-pound pail.
   a. True
   b. False
4. The drum bungs on a grease drum are of equal size.
   a. True
   b. False

5. Bulk storage is cleaner than drummed product and does not need additional filtration.
   a. True
   b. False
6. The use of antiwear hydraulic oil should be restricted to high-pressure hydraulic systems.
   a. True
   b. False
7. Grease in tubes should not be used in an operation because of their cost.
   a. True
   b. False
8. High-pressure low-volume pumps should be used on large bulk oil tanks to ensure that the product can be easily transferred to smaller containers.
   a. True
   b. False

19
Lubricant Properties by Application

Selection

Lubricants are selected for certain applications by the particular needs of that application. Although many applications have similar needs, many require special consideration and properties in their lubricant. That is the reason there are multipurpose lubricants and lubricants designed for a particular application.

To fully understand the needs of a particular application allows the lubricant manufacturer and consumer to manufacture and select the proper lubricant for the specific application. Many of the lubricant's properties are common to many uses, such as friction reduction, oxidation stability, detergency, and antirust. In adding a particular feature to a product, such as an antioxidation additive, how high is up or how much is really needed to do the job and at what cost to the manufacturer and to the consumer? That is why it is important that the properties of a lubricant be fitted to an area of performance needed for a particular application, or phrased another way, the needs of an application are fulfilled by the abilities of the lubricant to do the job effectively.

Listed are common applications of lubricants and the properties considered desirable to satisfy the needs of those applications.

- *Reciprocating air compressors:* Stability at high temperatures, non-foaming, low deposit forming, compounding to combat wet conditions and rusting, correct viscosity to reduce deposits and reduce volatility.

- *Vane-type air compressors:* The correct viscosity for sealing around the vanes, oxidation stability for high temperatures, and freedom from deposits.

- *Automatic transmission fluid:* Correct viscosity and viscosity index, proper frictional characteristics, antiwear and rust, seal control, and excellent low-temperature fluidity.

- *Cutting oil:* Proper viscosity, proper amount of antiweld for the metal being worked, antimisting, minimized smoking, maximum tool life, lubrication and cooling, antirusting, controlled chemical activity, good color and odor, and good surface finish on the workpiece.

- *Broaching fluid:* High chemical activity, high viscosity to stay on the broach, good cooling and tool life, and reasonable odor and color.

- *Bearing oil:* Good oxidation resistance, antirust, correct viscosity for the application, no corrosive additives, and pour point compatible with the application.

- *Diesel and gasoline engine oil:* Detergent-dispersant additives designed for diesel or gasoline service, rust- and oxidation-resistant, high viscosity index, and/or the use of shear stable viscosity index improvers, antiscuff additives, and proper TBN for high-sulfur fuels.

- *Hydraulic fluid:* Quality stock oil, antiscuff additives, rust- and oxidation-inhibited, high viscosity index, nonfoaming, and the proper viscosity.

- *Automotive gear oil:* Extreme-pressure additives to satisfy latest API requirements, rust- and oxidation-inhibited, nonfoaming, proper viscosity, and shear stable VI improvers.

Now that we have seen the properties required by application, we should also consider properties required by conditions involved with the applications. They are just as important as is the application for successful lubrication.

Conditions	Requirements
High temperatures	Higher viscosity and oxidation inhibitors
Wet conditions	Rust inhibitors and compounding
Low temperatures	Pour-point depressants, lower viscosity

Painted reservoirs	Avoid some fire-resistant fluids
Seal leakage	Check compatibility of seals with fluid
Water contamination	To avoid emulsions, eliminate detergents
Fire hazards	Use fire-resistant fluids but check seals
Overheating at high speed	Lower viscosity of the fluid
Overheating at low speeds	Raise the viscosity of the fluid
Internal leakage	Raise the viscosity of the fluid until repaired

It is now easy to see why lubricant manufacturers have a wide variety of products packaged in a wide variety of package sizes. Everyone would like to be able to use just one oil and one grease to lubricate their entire operation but that is only possible if the operation does only one thing.

Every different application is different from the other, and then even the same product may be required in a different viscosity. There is not a universal application or product. There are, however, many products which can serve multiple functions. For instance, a quality antiwear hydraulic required in a precision machine tool can also be used in the hydraulics of a forklift, as a bearing oil, as a gear oil in a gear set not requiring extreme-pressure additives, and as an air-line oiler lube in the lighter viscosities. So rather than stocking five different products, one quality multipurpose oil will serve you in five different types of applications, one drum rather than five, with easier ordering procedures and less potential of application errors.

The consolidation of products in an operation requires two major areas of knowledge: (1) knowledge of applications and operating conditions and (2) knowledge of each type of lubricant involved in the potential consolidation. Without this knowledge or expert assistance, a very costly oversight could occur in lost production, premature machine repair, and lost profit.

When working with your lubricant supplier, you, as a member of the maintenance management team, should be supplied with a list of products that might be used in different applications in case of an emergency and prior to the delivery of the proper product. This listing should be held by the maintenance management team for emergency use only and not provided to the people actually doing the maintenance. The reason is simple. There is a possibility that some workers who had a list of emergency substitutes might tend to use the emergency substitute rather than going out of their way to obtain the proper product from the oil storage facility.

Self-Quiz

Mark your answers to the questions and compare them with the answers at the end of the book.

1. It is not cost-effective to buy a product with additives you will not need.
   a. True
   b. False
2. Oxidation inhibitors are useful only in high-pressure hydraulic systems.
   a. True
   b. False
3. Sulfur compounds are always corrosive in a cutting fluid.
   a. True
   b. False
4. If a bearing is overheating at high speeds, you should probably:
   a. Add an oil heavier in viscosity
   b. Add an oil lighter in viscosity
   c. Fill the cavity completely with oil

Appendix

API S Classifications

SA. Moderate service in older engines. An SA oil normally will have little or no additives and is not currently recommended by any engine manufacturer. This classification is obsolete.

SB. Minimum duty and minimum protection. An SB oil will normally contain antiscuff and rust and oxidation additives. It is not currently recommended by any engine manufacturer. This classification is obsolete.

SC. Mild detergent-dispersant engine oil recommended for engine warranty for 1964–1967 engines. An oil of this classification would be rare in the marketplace today. This classification is obsolete.

SD. Provides more protection than needed for 1968–1970 warranty for some engines. This classification is obsolete.

SE. Warranty engine oil in 1971–1972. Provides more protection from deposits, rust, wear, and oxidation. This classification is obsolete.

SF. Greatly increases engine protection for 1980 and later warranty. This oil may be used where SC, SD, and SE oils are recommended. Improved performance in oxidation resistance and antiwear. You will see this classification on some diesel engine oils which are also designed for severe diesel service.

SG. 1989 warranty for current gasoline engines. SG oils also meet the properties of the diesel classification CC and would replace SE and SF oils for gasoline engines. Provides excellent rust and corrosion engine protection.

API C Classifications

CA. For mild- to moderate-duty diesel service with high-quality fuel. Should not be used unless the manufacturer recommends this grade. This classification is obsolete.

CB. Same as above except for fuels which are not of the highest quality. Should not be used unless the manufacturer recommends this grade. This classification is obsolete.

CC. Introduced in 1961, it is for moderate duty but is not now recommended for any current engines except perhaps by Detroit Diesel for their large 149 engine. The normal CD recommendation for this engine is still current but the higher ash level is giving some users problems. Meets the obsolete MIL specification of MIL-L-2104B. This classification is obsolete.

CD. For severe service on naturally aspirated, turbocharged, or supercharged diesel engines with high-quality fuel. This classification will soon be obsolete.

CD II. Also meets CD but provides highly effective control over deposits and wear. This classification will soon be obsolete.

CE. Severe service typical of many turbocharged or supercharged high-performance diesel engines. Operation includes low speed–high load and high speed–high load conditions. Provides improved control of oil consumption, oil thickening, and piston assembly deposits and wear related to the performance of a CD oil.

Proposed CF and CF-2. These are new diesel classifications which, at this time, are still not accepted by the API. The proposed CF classification would pass the new Cat 1-K test plus the obsolete L-38 test. The proposed CF-2 would be a CF oil which also passes the Detroit Diesel 6V92TA test. These classifications may change or may never be approved. At this time, it is doubtful that they will be finalized.

CF-4. This is the latest diesel classification, adopted in December 1990. It includes the proposed CF classification and passes the Mack EO-K2 test and a newer and tighter Cummins NTC 400 test.

These diesel classifications will continue to change depending on new military specifications and engine manufacturer's changes and their oil performance testing.

SAE Viscosity Grades for Engine Oils*

SAE viscosity grade	Low-Temperature (°C) viscosities cranking cP max	Low temperature (°C) viscosities pumping† cP max with no yield stress	Viscosity ‡§ (cs at 100°C)	
			Min	Max
0W	3250 at −30	30,000 at −35	3.8	
5W	3500 at −25	30,000 at −30	3.8	
10W	3500 at −20	30,000 at −25	4.1	
15W	3500 at −15	30,000 at −20	5.6	
20W	4500 at −10	30,000 at −15	5.6	
25W	6000 at − 5	30,000 at −10	9.3	
20			5.6	< 9.3
30			9.3	<12.5
40			12.5	<16.3
50			16.3	<21.9
60			21.9	<26.1

Reprinted with permission from SAE J300© February 1991, Society of Automotive Engineers, Inc.
1 cP = 1mPa; 1 cSt=1mm²/s.
*All values are critical specifications as defined by ASTM D 3244.
†ASTM D4684: Note that the presence of any yield stress detectable by this method constitutes a failure regardless of viscosity.
‡ASTM D 445.
§Some engine manufacturers also recommend limits on viscosity measured at 150°C and 10^6 S^{-1}.

SAE Viscosity Grades for Axle and Manual Transmission

SAE viscosity grade	Max temperature for viscosity of 150,000 cP*°C	Viscosity at 100°C† cSt	
		Min	Max
70W	−55	4.1	
75W	−40	4.1	
80W	−26	7.0	
85W	−12	11.0	
90		13.5	<24.0
140		24.0	<41.0
250		41.0	

Reprinted with permission from SAE J306© MAR 85, Society of Automotive Engineers, Inc.

*Centipoise (cP) is the customary absolute viscosity unit and is numerically equal to the corresponding SI unit of millipascal-second (mPa-s)

†Centistokes (cSt) is the customary kinematic viscosity unit and is numerically equal to the corresponding SI unit of square millimeter per second (mm^2/s).

The precision of ASTM Method D 2983 has not been established for determinations made at temperatures below −40°C; consequently, this fact should be realized in any producer-consumer relationship. It is expected that ASTM will shortly undertake work in the range down to −55°C for D2983.

ISO/ASTM Viscosity Grades for Industrial Fluid Lubricants

Viscosity grades	Centistokes at 40°C	Saybolt Universal at 100°F approx
2	1.98–2.42	33.0–34.5
3	2.88–3.52	36.0–38.0
5	4.14–5.06	40.5–43.5
7	6.12–7.48	47.5–52.0
10	9.00–11.0	58.0–65.0
15	13.5–16.5	76.0–88.0
22	19.8–24.2	105–125
32	28.8–35.2	150–180
46	41.4–50.6	215–260
68	61.2–74.8	320–380
100	90–110	470–575
150	135–165	710–870
220	198–242	1050–1250
320	288–352	1550–1850
460	414–506	2250–2700
680	612–748	3350–4000
1000	900–1100	4900–6000
1500	1350–1640	6800–7500

AGMA Standard Specifications, Lubrication of Industrial Enclosed Gear Drives

Rust- and oxidation-inhibited gear oils, AGMA lubricant No.	Viscosity range, ASTM system*		Extreme-pressure gear lubricants, AGMA lubricant No.
	SUS at 100°F	cSt at 37.8°C	
1	193–235	41.4–50.6	
2	284–347	61.2–74.8	2 EP
3	416–510	90–110	3 EP
4	626–765	135–165	4 EP
5	918–1122	198–242	5 EP
6	1335–1632	288–352	6 EP
7 comp†	1919–2346	414–505	7 EP
8 comp†	2837–3467	612–748	8 EP
8A comp†	4171–5098	900–1100	

Reproduced with the express consent of the American Gear Manufacturers' Association.
*Viscosity System for Industrial Lubricants, ASTM D2422. AGMA lubricant number viscosity ranges are identical to the ASTM system.
†Oils marked "comp" are compounded with fatty or synthetic oils.

Classification of Greases by NLGI Consistency Numbers

NLGI number	ASTM worked penetration
000	445–475
00	400–430
0	355–385
1	310–340
2	265–295
3	220–250
4	175–205
5	130–160
6	85–115

Reproduced with the express permission of the National Lubricating Grease Institute, Kansas City, Mo.

Machinability Tables, AISI

Resulfurized Metal		Leaded (Cont.)		Carbon (Cont.)		Carbon (Cont.)		Alloy steels (Cont.)		Alloy steels (Cont.)	
Metal	% Mach.	Metal	% Mach.	Metal	% Mach.	Metal	% Mach.	Metal	% Mach.	Metal	% Mach.
1108	81	12L14	170	1039	64	1541	57	4620	66	8640	66
1109	81	12L15	170	1040	64	1551	54	4621	66	8642	66
1110	81	41L40	77	1042	64	1552	49	4626	60	8645	64
1115	81	41L50	70	1043	57	1561	51	4718	60	8655	57
1116	94	10L18	92	1044	57	1566	49	4720	60	8720	66
1117	91	10L45	66	1045	57	1572	49	4815	51	8740	66
1118	91	10L50	60	1046	57	*Alloy steels*		4817	49	8822	64
1119	100	86L20	77	1048	54	1330	60	4820	49	9255	54
1120	81	*Carbon*		1049	54	1335	60	5015	78	9620	51
1125	81	1008	66	1050	54	1340	57	5060	76		
1132	81	1010	72	1053	54	1345	57	5120	57	*Stainless*	
1137	76	1012	72	1054	54	4012	78	5130	72	410	54
1138	72	1015	72	1055	51	4023	78	5132	72	416	110
1139	76	1016	78	1060	51	4024	78	5135	72	420	45
1140	76	1017	72	1059	51	4027	66	5140	70	420F	63
1141	72	1018	78	1064	49	4028	72	5145	66	430	54
1144	70	1019	78	1065	49	4037	72	5147	66	430F	91
1145	76	1020	72	1069	49	4047	66	5150	64	440A	45
1146	66	1021	78	1070	49	4118	78	5155	60	440B	42
1151	70	1022	78	1071	49	4130	72	5160	60	440C	40
1211	94	1023	76	1074	45	4137	70	E51100	40	440F	54
1212	100	1025	72	1075	45	4140	66	E52100	40	302	45
1213	136	1026	78	1078	45	4142	66	6118	66	303	78
1215	136	1029	70	1080	42	4145	64	6150	60	304	45
Leaded		1030	70	1084	42	4147	64	8615	70	316	45
11L17	104	1031	70	1085	42	4150	60	8617	66	321	36
11L37	84	1033	70	1086	42	4161	60	8620	66	347	36
11L41	79	1034	70	1090	42	4320	60	8622	66		
11L44	87	1035	70	1095	42	4340	57	8625	64		
12L13	170	1037	70	1524	66	4419	78	8626	64		
		1038	64	1527	66	4615	66	8630	72		
				1536	64			8637	70		

Glossary

The following definitions are simplified for ease of learning. Other sources give more complete definitions, but these allow you to know what a term means in the world of lubrication. There is no such thing as a complete listing of terms.

Additives: Chemicals or components added to fuels, oils, and greases to enhance performance in particular areas.

AGMA: American Gear Manufacturer's Association. A group of gear manufacturers, such as Falk, who have set viscosity standards for industrial gear oils including straight mineral oils (AGMA 5), extreme-pressure gear oils (AGMA 5 EP), and compounded gear oils containing fat (AGMA 7 comp). A table in the Appendix illustrates those AGMA viscosity grades.

Air Entrainment: The incorporation of air in the form of small bubbles dispersed in a fluid. An opaque appearance. Common when an excessive amount of defoamant is added to reduce foaming.

Ambient Temperature: The room air temperature surrounding the point of application.

Anhydrous: Free of water, especially water of crystallization.

Antiwear: An additive, normally zinc dithiophosphate, which is added to a hydraulic oil to prevent scuffing of the moving parts. This same additive is used in heavy-duty motor oils to prevent scuffing of the cam lobs. It is sometimes shown in a product name to denote it contains the additive (i.e., Mobil Hydraulic Oil AW). However, in the Mobil DTE or Shell Tellus series oils, which contain the antiwear additive, AW is not used in the product name.

API: American Petroleum Institute. A group of companies in petroleum and related fields.

API Gravity: A gravity scale arbitrarily related to specific gravity. An API gravity for water is 10. Lighter than water is more than 10 and heavier than water is less than 10.

Aromatic: Derived from, or characterized by, the presence of the benzene ring.

Autoignition: The temperature where an oil will ignite without a source of ignition.

AW: See Antiwear.

Barrel: A unit of liquid measure pertaining to 42 gallons of crude oil, gasoline, or fuel oils.

Bright Stock: Refined, high-viscosity lubricating oils usually made from residual stocks by suitable treatment.

Chain Stoppers: See Oxidation Inhibitors. This additive interrupts the chain reaction between oxygen and hydrocarbons to prevent or slow the formation of acidic materials and sludge caused by oxidation of the product.

Channel Point: See Pour Point. As you reduce the temperature of an oil below the pour point, you reach a point where you can run your finger through an oil and it will not fill in the trench you leave in the oil. Called the channel point, this is seen only in extremely cold climates. Example: The gearing in the rear end of a car. Although the gears might move, the gear oil will not flow back into the gear to lubricate it readily.

Cleveland Open Cup: Apparatus used to determine the flash and fire point of all petroleum products above 175°F except fuel oil.

Compounding: The addition of fatty oils and similar materials to lubricants to import special properties.

Defoamant: A very small amount is added to many different products which allows air in the oil to escape more rapidly. All oils will "foam," but an oil that does not release the air is said to "foam."

Detergent-Dispersant: Typically a combination of additives added to motor oils, designed to keep small particles separated so they do not combine and drop out in the engine. They will also tend to break up and keep separated those particles which are already in the engine.

Diluent: A solvent or other fluid which is combined with a very viscous product (like heavy oil or asphaltic-type material) for ease of application. A diluent is used to thin many open-gear lubricants so they can be applied with a brush or paddle. The diluent will then evaporate off, leaving the heavy or viscous material on the point of application.

Dropping Point: The temperature at which a grease passes from a semisolid to a liquid state under specified test conditions.

Drum: A round metal container which holds 55 gallons of oil or approximately 400 pounds of grease-type product. There are also half-size drums which hold approximately 30 gallons of oil.

Emulsifier: A substance used to promote or aid the emulsification of two liquids and to enhance the stability of the emulsion.

Emulsion: A mixture of two liquids which are not miscible with each other, i.e., oil-in-water cutting fluid. Water-in-oil is classified as an inverted emulsion.

Engler Viscosity: Another method of measuring the viscosity of an oil, primarily used in Europe.

EP, or Extreme Pressure: An additive, normally a sulfur-phosphorus combination, used to protect the gear teeth in an automotive or industrial gear set. The same type of additive can be used in grease or oil to produce the same protection, although the amount of additive will vary with the product.

Fillers: A term normally used to denote something nonchemical added to an oil or grease, i.e., moly, graphite, or zinc oxide.

Fire Point: See Flash Point. The temperature at which an oil, when subject to a source of ignition or a flame, will ignite and continue to burn. Typically the fire point is about 50°F above the flash point.

Flash Point: The temperature at which an oil, when subject to a flame, will flash or "poof." The vapors will ignite and then go out.

Floc Point: The temperature at which wax or solids separate as a definite floc.

Fluid Friction: A liquid's internal resistance to flow.

Foam: An agglomeration of gas bubbles separated from each other by a thin liquid film. If an oil is said to not foam, the small air bubbles will quickly combine, become larger bubbles, and then break and vent to the atmosphere. If this action occurs slowly, the oil is said to foam.

Four-Ball Tester: Machines used to evaluate a lubricant's antiwear qualities, frictional characteristics, or load-carrying capabilities. There are four steel ½-inch balls. Three of the balls are held together in a cup filled with lubricant while the fourth ball is rotated against them.

Friction: The resistance to motion.

FZG Test: A German gear test for evaluating EP properties. Use in the United States is becoming more common. Results will tend to demonstrate the real performance of a gear oil.

Grease: An oil or other lubricating fluid thickened with a soap or similar material.

Hypoid Gears: Gears in which the pinion axis intersects the plane of the ring gear at a point below the ring gear axis and above the outer edge of the ring gear, or above the ring gear axle and below the outer edge of the ring gear.

Inverted Emulsion: See Emulsion. A water-in-oil mixture. A drop of water surrounded by oil as in fire-resistant hydraulic fluids.

ISO: International Standard Organization, which has set viscosity standards for industrial oils. Viscosity is measured in centistokes at Celsius temperatures. (Example: cSt at 40°C.) This method is know as "kinematic viscosity."

Keg: A container which would typically hold 16 gallons of oil or approximately 120 pounds of grease-type product. Also called a "quarter drum."

Kinematic: See ISO.

Low Pour Point: Means an oil which will pour at low temperatures.

Metal Deactivators: An additive which prevents the oxidation-increasing catalytic effect of certain metals on the oil in the lubrication systems. Copper, lead, and iron are the most common.

Multiviscosity or Multigrade: A mineral oil to which a viscosity index improver has been added to reduce the thinning effect of heat. An SAE 10W-30 or 10W-40 is an oil which would meet the SAE 10W specification at 0°F (that is what the W means) and meets the SAE 30 or 40 specification at 210°F. A multiviscosity oil can *not* be a SAE 30-40 because that would mean it had two different viscosities at 210°F. Fluids with very high natural viscosity indexes, such as some synthetic fluids, are also classified as multiviscosity.

NLGI: The National Lubricating Grease Institute, an industry group that monitors grease and sets penetration standards for grading greases. See Penetration.

Oxidation: The connection of the oxygen molecule to the structure of the base fluid or material. Oil in a drum would oxidize very slowly. Oil in service, being mixed with air and circulated, would oxidize more rapidly. Rust is the slow oxidation of iron, fire is the fairly rapid oxidation of a material. An explosion is very rapid oxidation of a material.

Oxidation Inhibitors: See Chain Stoppers and Metal Deactivators. Some oxidation inhibitors perform both the chain stopper and metal deactivator functions. Basically, this additive reduces the opportunities for the oxygen molecule to grab onto the basic hydrogen carbon molecules of the petroleum product. This action increases the oxidation or service life of a petroleum product. An oil that is R&O means it contains *rust* and *oxidation* inhibitors.

Penetration: A test in which a cone is dropped into grease to measure the penetration or how hard or soft the grease is at room temperatures. The cone penetrates farther in a soft grease and therefore has a higher penetration number. The higher the penetration, the softer the grease. This penetration relates to an NLGI (National Lubricating Grease Institute) number. A number 0 grease is called an NLGI 0 grade and will be "softer" than an NLGI 1 or 2 grade.

Pour Point: The point (temperaturewise) where an oil will just move. An oil is reduced in temperature until it will not move and then 5°F is added, and that is the "pour point."

Pour-Point Depressant: An additive which prevents wax crystals from combining so an oil or fuel (diesel) will pour at a lower temperature.

RBOT Test: Rotary bomb oxidation test, or RBOT. This is a special fast test to determine the relative oxidation life of a new or used oil. The test results do not relate to any other test results, only to previous RBOT test results.

Rust: The slow oxidation of metal.

Saybolt, Saybolt Universal Seconds, SUS or SSU: A common method of measuring viscosity in the United States for years. Saybolt is the name of the method, Universal means it uses a given orifice, and Seconds means the time the given amount of oil takes to flow through the orifice. This method, although very common, is being replaced by the kinematic method.

SAE: Society of Automotive Engineers. An industry group which has set the viscosity brackets for motor oil grades and automotive gear oil grades (i.e., SAE 30, 15W-40, and 80W-90) When a number such as 10 is followed by a W, it means that specification is measured at 0°F. A table in the Appendix illustrates those SAE grades.

Thickener: The metallic soap or other material used to combine with oil or other lubricating fluid to make a grease.

Timken OK Load: This was a test, developed in the 1930s, to determine if an automotive gear oil *did* or *did not* contain extreme-pressure additives to handle the differentials or hypoid gear rear ends in a passenger car. It was never really designed to "measure" the amount or capabilities of an oil to handle extreme pressures. It is, however, basically used for that purpose today. You can get a high reading or fool it just by adding fat to the oil. There are several other test rigs to measure the same thing and there is little comparison between the test results.

TOST Test: Turbine oxidation stability test, or TOST. This is a measure of an oil's resistance to oxidation in hours. The test conditions include the introduction of oxygen and copper catalysts and do not directly relate to normal operating conditions of most oils. A cheap oil, with enough oxidation additives, can pass a 2000-hour TOST test and then completely fall apart. A better oil, with oxidation additive, can just pass the test and after the depletion of the oxidation inhibitors can continue to perform based on the good oxidation resistance of the base oil. The difference in performance is the quality of the base stock oil.

Viscosity: The measurement of the resistance to flow of a fluid. The higher the viscosity, the more resistance to flow. A "heavy" oil has more resistance to flow than a "light" oil. Normally measured at 100°F and 210°F for Saybolt Universal and 40°C and 100°C for other classifications.

Viscosity Index: The rate at which an oil resists thinning with temperatures. An oil with a VI of 150 resists thinning better than an oil with a VI of 100. A multigrade motor oil resists thinning better than a monograde motor oil as temperatures increase. Commonly called "VI."

Viscosity Index Improver: We know what viscosity and viscosity index means. Viscosity index improver means an additive to "improve or increase the viscosity index." A VI improver *increases* an oil's resistance to thinning as it is heated. It is commonly used in multiviscosity or multigrade motor oils. Since a VI improver increases the viscosity as well as the viscosity index, it must be taken into consideration when formulating an oil. (Example: Taking an oil in the SAE 10 range, adding a VI improver, would give us some oil like an SAE 20W-40. Adding the VI improver to an SAE 30 would give us an SAE 40-something. See Multiviscosity to see how an oil can be an SAE 10W and an SAE 40.

Self-Quiz Answers

						Questions							
Chapter	1	2	3	4	5	6	7	8	9	10	11	12	13
2	a	b	a	b	a	c							
3	d	c	a	b	c	a	d						
4	b	b	b	b	a	b	b						
5	b	b	c	a	b	b	c						
6	b	b	c	b	a	a	b						
7	a	b	b	b	b								
8	b	b	b	b	a	b	c						
9	b	a	c	b	b	b	a						
10	*	†	c	b	b	b	b	b	c	b	b	b	a
11	b	b	b	c	c	b	b						
12	a	a	b	b	b	b	b						
13	‡	b	a	a	b								
14	b	b	b	b	b								
15	No quiz												
16	b	a	b	b	b	b	a						
17	No quiz												
18	b	b	b	b	b	b	b	b					
19	b	b	b	c									

*Gear, piston, and vane.
†Rust and oxidation.
‡Reciprocating, centrifugal, axial-flow, and rotary.

Index

About The Author

Robert W. Miller is a consultant with more than 34 years' experience with Shell Oil Company and Mobil Oil in the engineering and marketing of petroleum products to a broad range of industries. Regarded as an expert in the field of industrial lubrication, he has lectured frequently on the subject at Arizona State University. Mr. Miller resides in Tempe, Arizona.